Target

Get back on track

GRADE 5

Edexcel GCSE (9-1)
Mathematics
Shape and Statistics

Diane Oliver

Pearson

Published by Pearson Education Limited, 80 Strand, London, WC2R ORL.

www.pearsonschoolsandfecolleges.co.uk

Text © Pearson Education Limited 2016
Typeset by Tech-Set Ltd, Gateshead
Original illustrations © Pearson Education Ltd 2016

The right of Diane Oliver to be identified as author of this work has been asserted by her in accordance with the Copyright, Designs and Patents Act 1988.

First published 2016

20
10 9 8 7 6

British Library Cataloguing in Publication Data
A catalogue record for this book is available from the British Library

ISBN 978 0 435 18334 9

Printed in Italy by Lego S.p.A

Helping you to formulate grade predictions, apply interventions and track progress.

Any reference to indicative grades in the Pearson Target Workbooks and Pearson Progression Services is not to be used as an accurate indicator of how a student will be awarded a grade for their GCSE exams.

You have told us that mapping the Steps from the Pearson Progression Maps to indicative grades will make it simpler for you to accumulate the evidence to formulate your own grade predictions, apply any interventions and track student progress.

We're really excited about this work and its potential for helping teachers and students. It is, however, important to understand that this mapping is for guidance only to support teachers' own predictions of progress and is not an accurate predictor of grades.

Our Pearson Progression Scale is criterion referenced. If a student can perform a task or demonstrate a skill, we say they are working at a certain Step according to the criteria. Teachers can mark assessments and issue results with reference to these criteria which do not depend on the wider cohort in any given year. For GCSE exams however, all Awarding Organisations set the grade boundaries with reference to the strength of the cohort in any given year. For more information about how this works please visit: https://qualifications.pearson.com/en/support/support-topics/results-certification/understanding-marks-and-grades.html/Teacher

Each practice question features a Step icon which denotes the level of challenge aligned to the Pearson Progression Map and Scale.

To find out more about the Progression Scale for Maths and to see how it relates to indicative GCSE 9–1 grades go to www.pearsonschools.co.uk/ProgressionServices

Contents

Useful formulae

Unit 1 Circles

Circumference of a circle $(C) = \pi d = 2\pi r$

Arc length $= \dfrac{x}{360} \times 2\pi r$

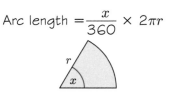

Area of a circle $= \pi r^2$

Area of a sector $= \dfrac{x}{360} \times \pi r^2$

Unit 2 Volume and surface area

Volume of a cuboid $(V) = lwh$

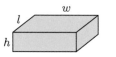

Surface area of a cuboid $= 2lw + 2hw + 2lh$

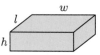

Volume of a cylinder $(V) = \pi r^2 h$

Surface area of a cylinder $= 2\pi r^2 + 2\pi rh$

Volume of a sphere $(V) = \frac{4}{3}\pi r^3$

Surface area of a sphere $= 4\pi r^2$

Volume of a cone $(V) = \frac{1}{3}\pi r^2 h$

Surface area of a cone $= \pi rl + \pi r^2$

Volume of a pyramid $(V) = \frac{1}{3}bh$

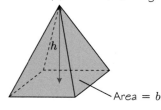

Area $= b$

Unit 7 Probability

Probability $= \dfrac{\text{number of successful outcomes}}{\text{total number of possible outcomes}}$

Unit 10 Percentages

Percentage change $= \dfrac{\text{actual change}}{\text{original amount}} \times 100$

Glossary

Unit 1 Circles

The **circumference** is the perimeter of a circle.

The **radius** is the distance from the centre to the circumference.

The **diameter** is a line from one side of the circle to the other through the centre.

An **arc** is part of the circumference.

A **sector** is a slice of a circle between an arc and two radii.

A **formula** is a general rule that shows a relationship between variables.

Substitute means to put numbers in place of letters.

Unit 2 Volume and surface area

A **net** is a 2D shape that folds up to make a 3D solid.

A **prism** is a solid with the same cross-section throughout its length.

The **volume** of a 3D solid is the amount of space inside it.

The **surface area** of a 3D solid is the total area of all its faces.

A **cuboid** is a solid in which each of the faces is a rectangle.

A **pyramid** is a solid which has triangular faces that converge to a point at the top. The base is often a rectangle or triangle, but it could be a pentagon, hexagon or any other polygon.

Unit 3 Angles

A **polygon** is a 2D shape bounded by straight edges.

A **regular** polygon has all equal side lengths and all equal angles.

An **interior angle** is an angle inside a polygon.

An **exterior angle** is the angle between an edge of a polygon and a line extended from the neighbouring edge.

The **sum of the exterior angles** of a polygon is always 360°.

Angles in a triangle add up to 180°.

Angles on a straight line add up to 180°.

Unit 4 Vectors

A **vector** describes a translation. A vector can be written in several ways: as $\begin{pmatrix} a \\ b \end{pmatrix}$, $\overrightarrow{AB}$, or **a**.

Unit 5 Transformations

A **transformation** transforms a shape (the **object**) to a different position.

The **image** is the shape after a transformation. Two shapes are **similar** when corresponding sides are in the same ratio and corresponding angles are equal. Similar shapes may be different sizes.

A **reflection** is a type of transformation. All points on the image are the same distance from the **line of reflection** as the points on the object, but on the opposite side.

A **rotation** is a type of transformation. You rotate a shape by turning it around a point, called the **centre of rotation**. To describe the rotation you need to specify the centre, the angle and the direction.

To **enlarge** a shape you multiply all the side lengths by the same number.

The number that the side lengths are multiplied by in an enlargement is called the **scale factor**.

The **centre of enlargement** is the point from which a shape has been enlarged.

A **translation** of a 2D shape is a slide across a flat surface. To describe a translation you need to give the horizontal and vertical movements.

A **vertex** of a shape is a point where two sides join.

Unit 6 Averages

The **mean** of a set of data is the total of all the values divided by the number of values.

The **median** is the middle value when the values are written in order.

The **mode** is the most common value in a set of data.

The **range** is the difference between the smallest and largest data values.

The **average** of a set of data gives a typical value for the data. The mode, median and mean are different ways of describing the average of a set of data.

When the data is grouped, you can calculate the **estimated mean**.

A **class interval** for grouped data is a range of values, represented by divisions on a chart. Class intervals do not overlap.

The **midpoint** of a class interval is the middle value.

The **frequency** for grouped data is the number of times a value within the range of the class interval occurs.

Unit 7 Probability

A **factor** is a whole number that will divide exactly into another number.

A **prime number** has exactly two factors, itself and 1.

Probability is the likelihood that an **event** will happen.

A **Venn diagram** is a way of representing relationships between sets of data using circles or ovals to show the boundaries of the different sets.

A **tree diagram** shows two or more events and their probabilities.

If one event does not affect the outcome of another event, the two events are **independent events**.

If one event depends on the outcome of another event, the two events are **dependent events**.

Unit 8 Scatter graphs

A **scatter graph** displays the relationship between two sets of data. The sets of data displayed are called **variables**.

The relationship between two sets of data is called **correlation**. The two sets of data can display positive, negative or no correlation.

A **line of best fit** is a straight line drawn through the middle of the points on a scatter graph. It should pass as near to as many points as possible and represent the **trend** of the points.

Interpolation is using the line of best fit to predict data values within the range of the given data.

Extrapolation is using the line of best fit to predict data values outside the range of the given data.

In a **causal** relationship between two variables, changes to one variable cause changes to the other. Two sets of data can be correlated without there being a causal relationship.

Unit 9 Sequences

A **number sequence** is a pattern of numbers that follow a **rule**. The numbers in a sequence are called **terms**.

The **nth term** or **general term** of a sequence is the rule for working out the term at position n.

Arithmetic sequences are sequences where the terms increase or decrease by the same number, called the **common difference**.

An **integer** is a positive or negative whole number, or zero.

Unit 10 Percentages

Percentage means 'out of 100'.

Compound interest is when interest is calculated on the initial amount and on any previous interest.

To **depreciate** means to decrease in value.

Per annum means per year.

(1) Circles

This unit will help you to calculate perimeters and areas of circles and sectors, and to find the radius given the circumference or area.

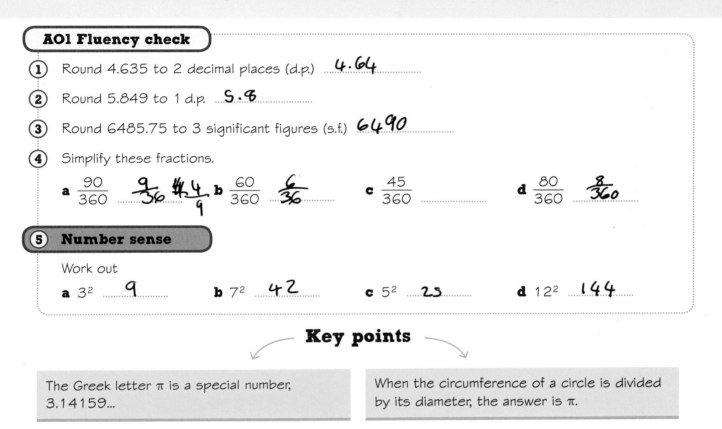

AO1 Fluency check

(1) Round 4.635 to 2 decimal places (d.p.) **4.64**

(2) Round 5.849 to 1 d.p. **5.8**

(3) Round 6485.75 to 3 significant figures (s.f.) **6490**

(4) Simplify these fractions.

a $\frac{90}{360}$ **$\frac{9}{36}$ $\frac{4}{9}$** b $\frac{60}{360}$ **$\frac{6}{36}$** c $\frac{45}{360}$ d $\frac{80}{360}$ **$\frac{8}{360}$**

(5) **Number sense**

Work out

a 3^2 **9** b 7^2 **42** c 5^2 **25** d 12^2 **144**

Key points

The Greek letter π is a special number, 3.14159...

When the circumference of a circle is divided by its diameter, the answer is π.

These **skills boosts** will help you to solve problems involving circles and sectors.

1 Circumference of a circle   2 Area of a circle   3 Arc length and perimeter of a sector   4 Area of a sector

You might have already done some work on circles. Before starting the first skills boost, rate your confidence with these questions.

(1) Work out the circumference of a circle with diameter 4 cm.

(2) Work out the area of a circle with radius 7 cm.

(3) Work out the perimeter of a quarter circle with radius 6 cm.

(4) Work out the area of a semicircle with diameter 10 cm.

How confident are you?

1 Circumference of a circle

Guided practice

Work out the circumference of the circle.
Give your answer to 2 d.p.

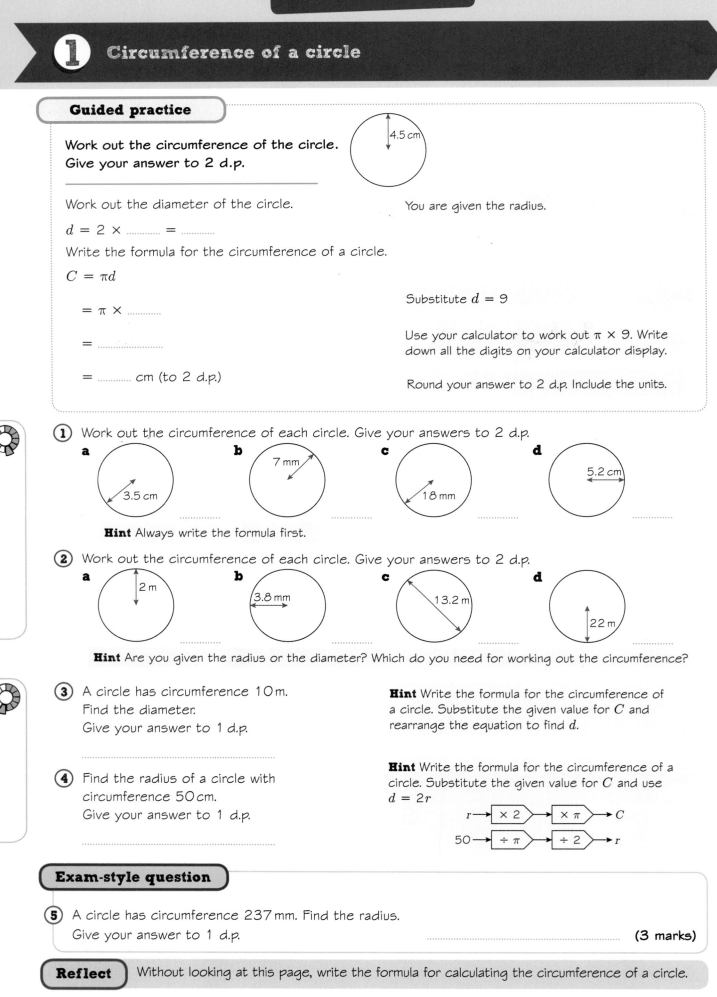

4.5 cm

Work out the diameter of the circle.

You are given the radius.

$d = 2 \times \text{............} = \text{............}$

Write the formula for the circumference of a circle.

$C = \pi d$

$= \pi \times \text{............}$

Substitute $d = 9$

$= \text{............................}$

Use your calculator to work out $\pi \times 9$. Write down all the digits on your calculator display.

$= \text{............}$ cm (to 2 d.p.)

Round your answer to 2 d.p. Include the units.

① Work out the circumference of each circle. Give your answers to 2 d.p.

a   3.5 cm

b   7 mm

c   18 mm

d   5.2 cm

Hint Always write the formula first.

② Work out the circumference of each circle. Give your answers to 2 d.p.

a   2 m

b   3.8 mm

c   13.2 m

d   22 m

Hint Are you given the radius or the diameter? Which do you need for working out the circumference?

③ A circle has circumference 10 m.
Find the diameter.
Give your answer to 1 d.p.

Hint Write the formula for the circumference of a circle. Substitute the given value for C and rearrange the equation to find d.

④ Find the radius of a circle with circumference 50 cm.
Give your answer to 1 d.p.

Hint Write the formula for the circumference of a circle. Substitute the given value for C and use $d = 2r$

$r \rightarrow \boxed{\times 2} \rightarrow \boxed{\times \pi} \rightarrow C$

$50 \rightarrow \boxed{\div \pi} \rightarrow \boxed{\div 2} \rightarrow r$

Exam-style question

⑤ A circle has circumference 237 mm. Find the radius.
Give your answer to 1 d.p. .... **(3 marks)**

Reflect Without looking at this page, write the formula for calculating the circumference of a circle.

2 Area of a circle

Guided practice

Work out the area of the circle. Give your answer to 1 d.p.

7 cm

Worked exam question

Write the formula for the area of a circle.

$A = \pi r^2$

$= \pi \times \underline{\hspace{1cm}}^2$

Substitute $r = 7$.

$= \underline{\hspace{2cm}}$

Use the π button on your calculator to work out $\pi \times 7^2$. Write down all the digits on your calculator display.

$= \underline{\hspace{1cm}}$ cm^2 (to 1 d.p.)

Round your answer to 1 d.p. Include the units.

① Work out the area of each circle. Give your answers to 1 d.p.

a 3 cm

b 7.5 mm

c 5 m

d 2.4 m

Hint The radius is given in mm, so the unit for the area is mm^2.

② Work out the area of each circle. Give your answers to 1 d.p.

a 26 cm

b 13 m

c 50 mm

Hint The formula for the area of a circle uses the radius.

③ Work out the area of each circle. Give your answers to 1 d.p.

a radius 4.5 cm b diameter 18 mm c radius 2.3 cm d diameter 37 mm

④ Find the radius of a circle with area 120 mm^2.
Give your answer to 1 d.p.

Hint Substitute $A = 120$ into the formula for the area of a circle then rearrange the equation to find the radius.

Exam-style question

⑤ A circle has area 3.67 m^2. Find the radius.
Give your answer to 2 d.p.

(3 marks)

Reflect Without looking at this page, write the formula for calculating the area of a circle.

3 Arc length and perimeter of a sector

The arc length of a sector is a fraction of the circumference of the circle.

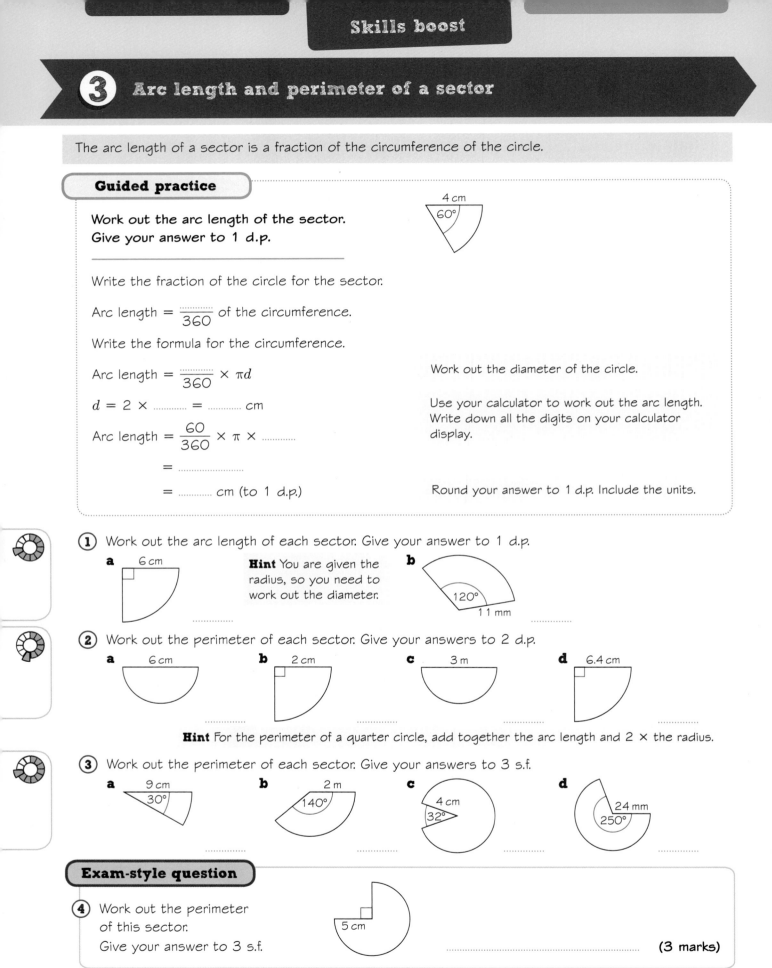

Guided practice

Work out the arc length of the sector.
Give your answer to 1 d.p.

4 cm
60°

_____

Write the fraction of the circle for the sector.

Arc length = $\dfrac{\text{..........}}{360}$ of the circumference.

Write the formula for the circumference.

Arc length = $\dfrac{\text{..........}}{360} \times \pi d$

$d = 2 \times$ .... = ... cm

Arc length = $\dfrac{60}{360} \times \pi \times$ ....

= ....

= .... cm (to 1 d.p.)

Work out the diameter of the circle.

Use your calculator to work out the arc length. Write down all the digits on your calculator display.

Round your answer to 1 d.p. Include the units.

---

① Work out the arc length of each sector. Give your answer to 1 d.p.

a 6 cm

Hint You are given the radius, so you need to work out the diameter.

b 120° 11 mm

---

② Work out the perimeter of each sector. Give your answers to 2 d.p.

a 6 cm **b** 2 cm **c** 3 m **d** 6.4 cm

Hint For the perimeter of a quarter circle, add together the arc length and 2 × the radius.

---

③ Work out the perimeter of each sector. Give your answers to 3 s.f.

a 9 cm 30°

b 2 m 140°

c 4 cm 32°

d 24 mm 250°

Exam-style question

④ Work out the perimeter
of this sector.
Give your answer to 3 s.f.

5 cm

.... (3 marks)

Reflect Did you simplify your fractions before finding the answer?
Does it affect your answer?

4 Area of a sector

The area of a sector is a fraction of the area of the circle.

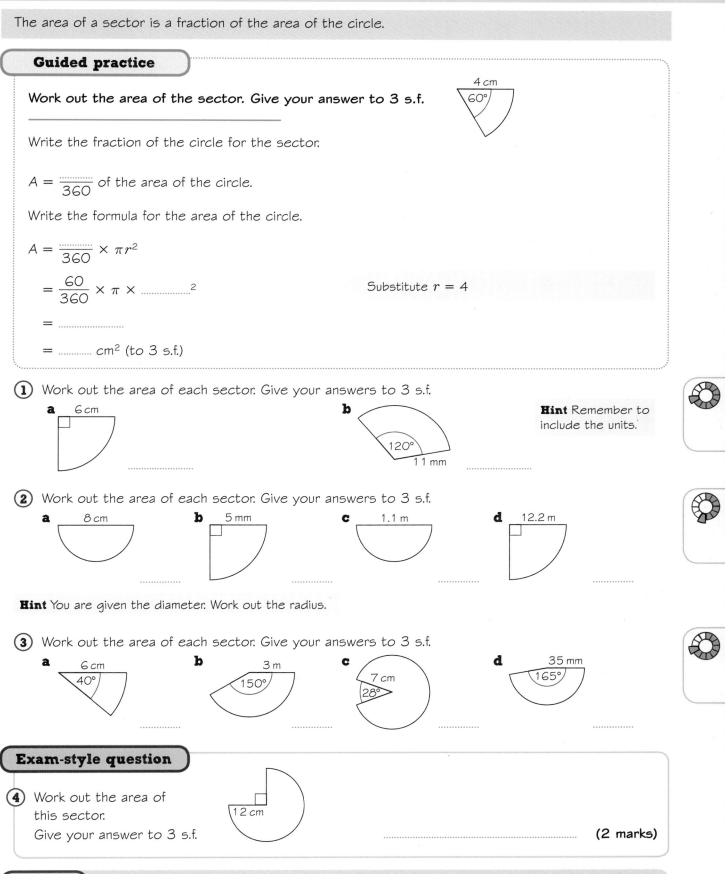

Guided practice

Work out the area of the sector. Give your answer to 3 s.f.

4 cm
60°

Write the fraction of the circle for the sector.

$A = \dfrac{.........}{360}$ of the area of the circle.

Write the formula for the area of the circle.

$A = \dfrac{.........}{360} \times \pi r^2$

$\quad = \dfrac{60}{360} \times \pi \times ....^2$ Substitute $r = 4$

$\quad = ...$

$\quad = ... \text{ cm}^2$ (to 3 s.f.)

① Work out the area of each sector. Give your answers to 3 s.f.

a 6 cm

b 120° 11 mm

Hint Remember to include the units.

② Work out the area of each sector. Give your answers to 3 s.f.

a 8 cm   **b** 5 mm   **c** 1.1 m   **d** 12.2 m

Hint You are given the diameter. Work out the radius.

③ Work out the area of each sector. Give your answers to 3 s.f.

a 6 cm 40°   **b** 3 m 150°   **c** 7 cm 28°   **d** 35 mm 165°

Exam-style question

④ Work out the area of this sector. Give your answer to 3 s.f.

12 cm

(2 marks)

Reflect   Without looking at this page, write the formula for calculating the area of a sector.

Practise the methods

Answer this question to check where to start.

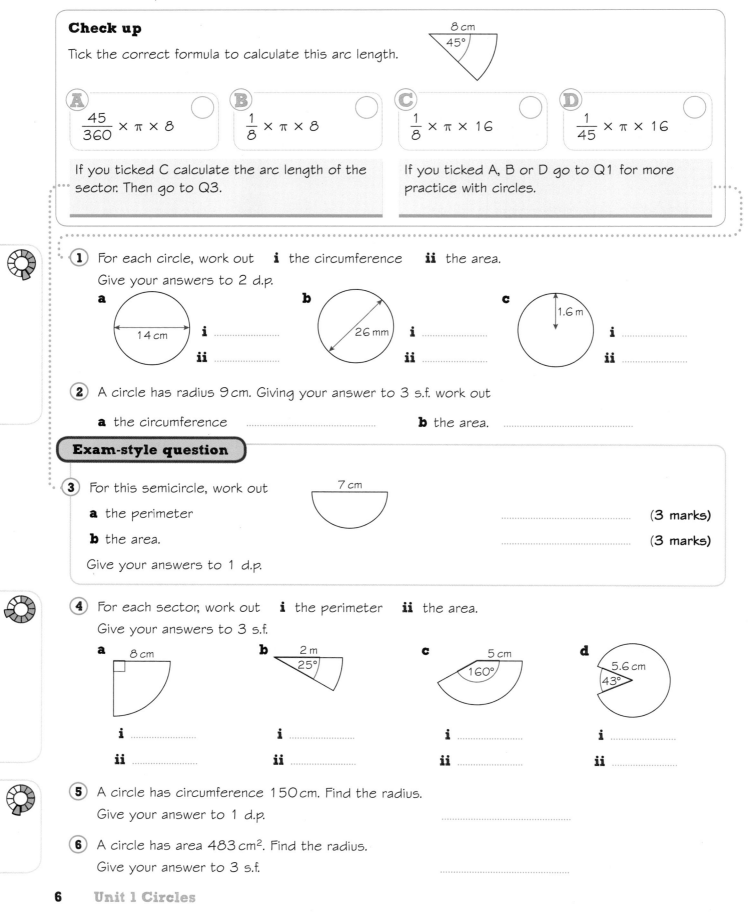

Check up

Tick the correct formula to calculate this arc length.

8 cm
45°

A $\frac{45}{360} \times \pi \times 8$ ◯

B $\frac{1}{8} \times \pi \times 8$ ◯

C $\frac{1}{8} \times \pi \times 16$ ◯

D $\frac{1}{45} \times \pi \times 16$ ◯

If you ticked C calculate the arc length of the sector. Then go to Q3.

If you ticked A, B or D go to Q1 for more practice with circles.

1. For each circle, work out **i** the circumference **ii** the area.
   Give your answers to 2 d.p.

   a 14 cm
   i ....
   ii ....

   b 26 mm
   i ....
   ii ....

   c 1.6 m
   i ....
   ii ....

2. A circle has radius 9 cm. Giving your answer to 3 s.f. work out

   a the circumference ....   **b** the area. ....

Exam-style question

3. For this semicircle, work out   7 cm

   a the perimeter .... (3 marks)

   b the area. .... (3 marks)

   Give your answers to 1 d.p.

4. For each sector, work out **i** the perimeter **ii** the area.
   Give your answers to 3 s.f.

   a 8 cm
   i ....
   ii ....

   b 2 m 25°
   i ....
   ii ....

   c 5 cm 160°
   i ....
   ii ....

   d 5.6 cm 43°
   i ....
   ii ....

5. A circle has circumference 150 cm. Find the radius.
   Give your answer to 1 d.p. ....

6. A circle has area 483 cm². Find the radius.
   Give your answer to 3 s.f. ....

Problem-solve!

Exam-style questions

1 Akram has a flower garden in the shape of a circle.

The diameter of the garden is 5 metres.

He wants to put fencing around the edge of the garden.

The fencing costs £3.40 per metre.

Work out the total cost of the fencing. .... (3 marks)

2 The diagram shows a circle drawn inside a square.

The circle has a radius of 5 cm.

The square has a side of length 10 cm.

Work out the area of the shaded region.

Give your answer to 1 d.p. .... (3 marks)

3 The diagram shows a slice of pizza.

The radius of the pizza is 18 cm.

The angle of the slice is 30°.

Work out the total area of four slices of pizza.

Give your answer correct to 3 s.f. .... (2 marks)

4 The diagram shows a semicircle drawn inside a rectangle.

The semicircle has a diameter of 12 cm.

The rectangle is 12 cm by 6 cm.

Work out the total area of the shaded regions.

Give your answer correct to 3 s.f. .... (4 marks)

5 Jamie is making a circular birthday cake.

He has a piece of ribbon 50 cm long to put around the cake.

Jamie has a 15 cm diameter cake tin
and an 18 cm diameter cake tin.

Which tin should he use? .... (3 marks)

6 A quarter circle has area 10 m². Work out the radius.

Give your answer to 2 d.p. .... (3 marks)

Now that you have completed this unit, how confident do you feel?

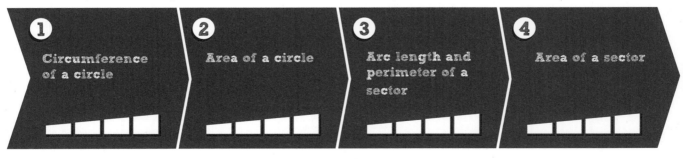

1 Circumference of a circle

2 Area of a circle

3 Arc length and perimeter of a sector

4 Area of a sector

② Volume and surface area

This unit will help you to calculate the surface area and volume of solids and to use volume to calculate the length of prisms.

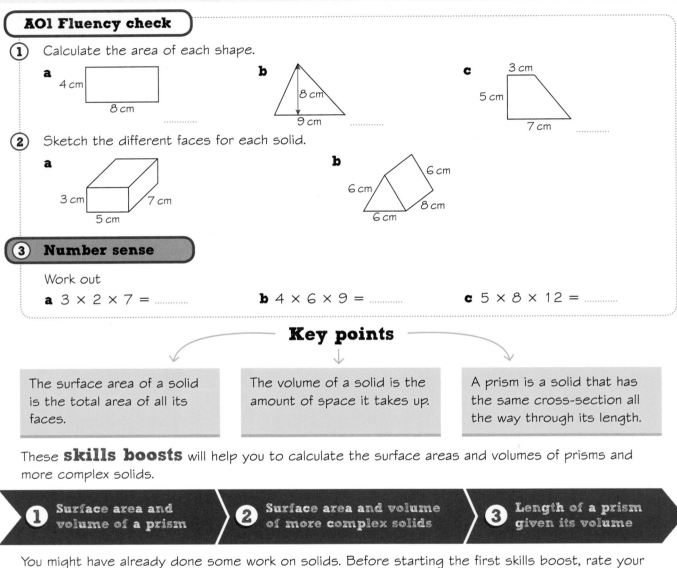

A01 Fluency check

① Calculate the area of each shape.

a 4 cm, 8 cm

b 8 cm, 9 cm

c 3 cm, 5 cm, 7 cm

② Sketch the different faces for each solid.

a 3 cm, 7 cm, 5 cm

b 6 cm, 6 cm, 6 cm, 8 cm

③ **Number sense**

Work out

a $3 \times 2 \times 7 = $ ....

b $4 \times 6 \times 9 = $ ....

c $5 \times 8 \times 12 = $ ....

Key points

| The surface area of a solid is the total area of all its faces. | The volume of a solid is the amount of space it takes up. | A prism is a solid that has the same cross-section all the way through its length. |

These **skills boosts** will help you to calculate the surface areas and volumes of prisms and more complex solids.

① Surface area and volume of a prism

② Surface area and volume of more complex solids

③ Length of a prism given its volume

You might have already done some work on solids. Before starting the first skills boost, rate your confidence using surface area and volume in different ways.

① Work out the surface area and volume of the cuboid. 3 cm, 6 cm, 4 cm

② Work out the volume of a square-based pyramid with a base 8 cm wide and a vertical height of 11 cm. Give your answer to the nearest whole number.

③ Rearrange the formula $V = \pi r^2 h$ to make h the subject.

How confident are you?

1 Surface area and volume of a prism

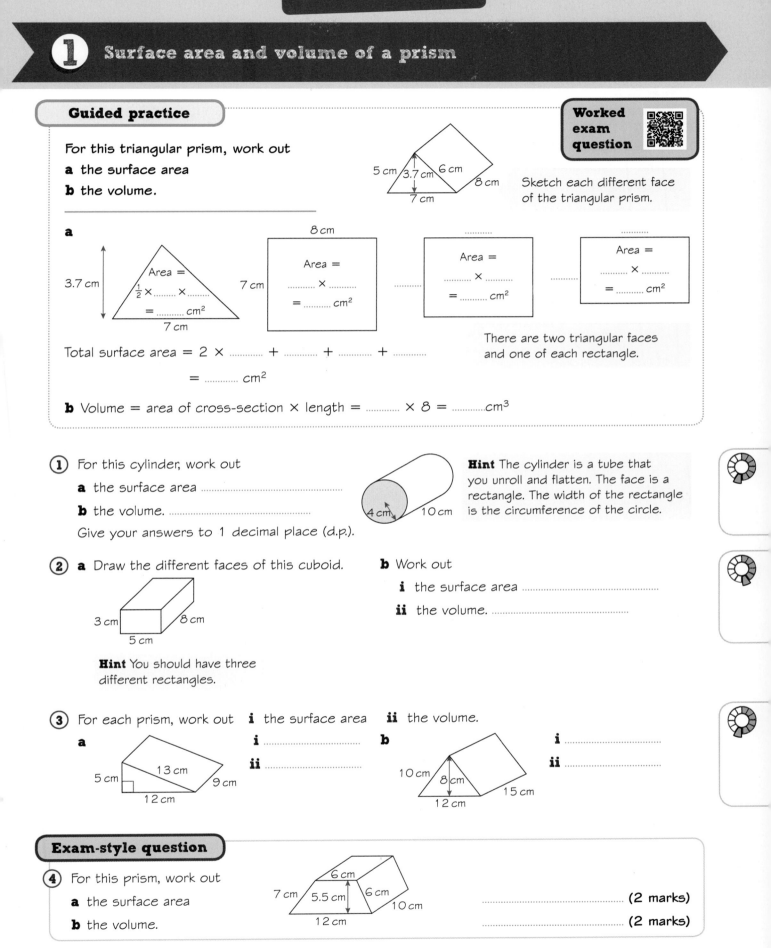

Guided practice

For this triangular prism, work out
a the surface area
b the volume.

Worked exam question

5 cm 3.7 cm 6 cm 8 cm 7 cm

Sketch each different face of the triangular prism.

a

Area =
$\frac{1}{2}$ × .... × ....
= .... cm²

3.7 cm   7 cm

8 cm

Area =
.... × ....
= .... cm²

7 cm

Area =
.... × ....
= .... cm²

Area =
.... × ....
= .... cm²

There are two triangular faces and one of each rectangle.

Total surface area = 2 × .... + ... + ... + ...

= .... cm²

b Volume = area of cross-section × length = ... × 8 = ... cm³

① For this cylinder, work out
  a the surface area ....
  b the volume. ....
  Give your answers to 1 decimal place (d.p.).

Hint The cylinder is a tube that you unroll and flatten. The face is a rectangle. The width of the rectangle is the circumference of the circle.

4 cm   10 cm

② **a** Draw the different faces of this cuboid.

3 cm 8 cm 5 cm

b Work out
  i the surface area ....
  ii the volume. ....

Hint You should have three different rectangles.

③ For each prism, work out **i** the surface area   **ii** the volume.
  a

5 cm 13 cm 9 cm 12 cm

  i ....
  ii ....

  b

10 cm 8 cm 15 cm 12 cm

  i ....
  ii ....

Exam-style question

④ For this prism, work out
  a the surface area
  b the volume.

6 cm 7 cm 5.5 cm 6 cm 10 cm 12 cm

.... **(2 marks)**
.... **(2 marks)**

Reflect Without looking at this page, write the formula for working out the volume of a prism.

2 Surface area and volume of more complex solids

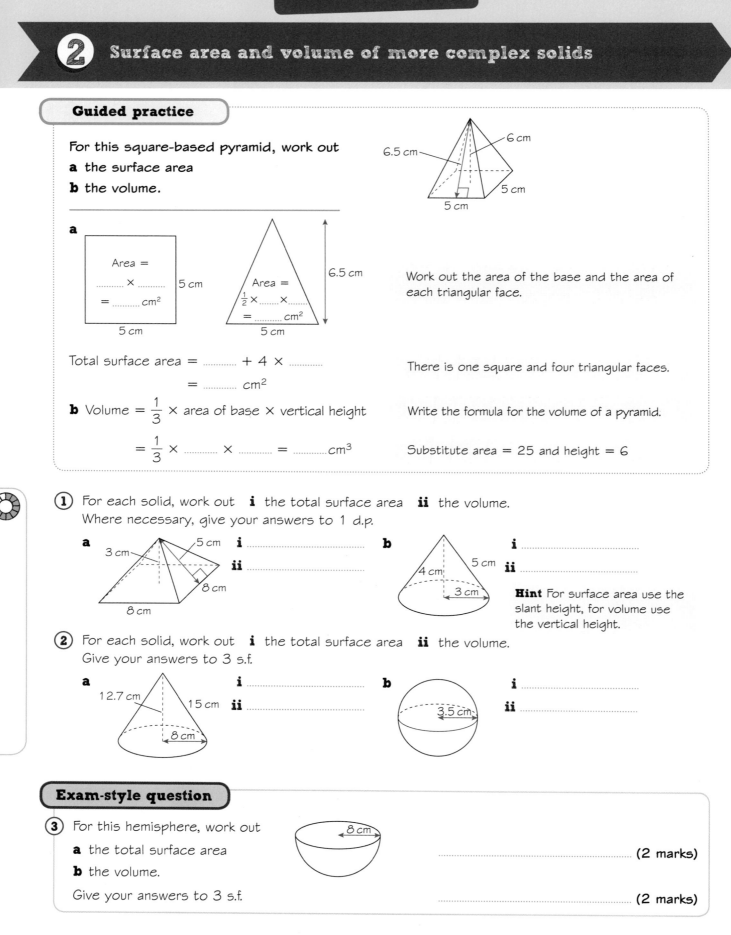

Guided practice

For this square-based pyramid, work out
a the surface area
b the volume.

6.5 cm   6 cm   5 cm   5 cm

a

Area =
.... × ....
= .... cm²

5 cm   5 cm

Area =
$\frac{1}{2}$ × .... × ....
= .... cm²

6.5 cm   5 cm

Work out the area of the base and the area of each triangular face.

Total surface area = ... + 4 × ....
= .... cm²

There is one square and four triangular faces.

b Volume = $\frac{1}{3}$ × area of base × vertical height

Write the formula for the volume of a pyramid.

= $\frac{1}{3}$ × .... × .... = ...cm³

Substitute area = 25 and height = 6

① For each solid, work out **i** the total surface area **ii** the volume.
Where necessary, give your answers to 1 d.p.

a 3 cm 5 cm 8 cm 8 cm

i ....
ii ....

b 5 cm 4 cm 3 cm

i ....
ii ....

Hint For surface area use the slant height, for volume use the vertical height.

② For each solid, work out **i** the total surface area **ii** the volume.
Give your answers to 3 s.f.

a 12.7 cm 15 cm 8 cm

i ....
ii ....

b 3.5 cm

i ....
ii ....

Exam-style question

③ For this hemisphere, work out
  a the total surface area
  b the volume.
  Give your answers to 3 s.f.

8 cm

.... (2 marks)

.... (2 marks)

Reflect   What strategies did you use to help you work out the surface area of these solids?

3 Length of a prism given its volume

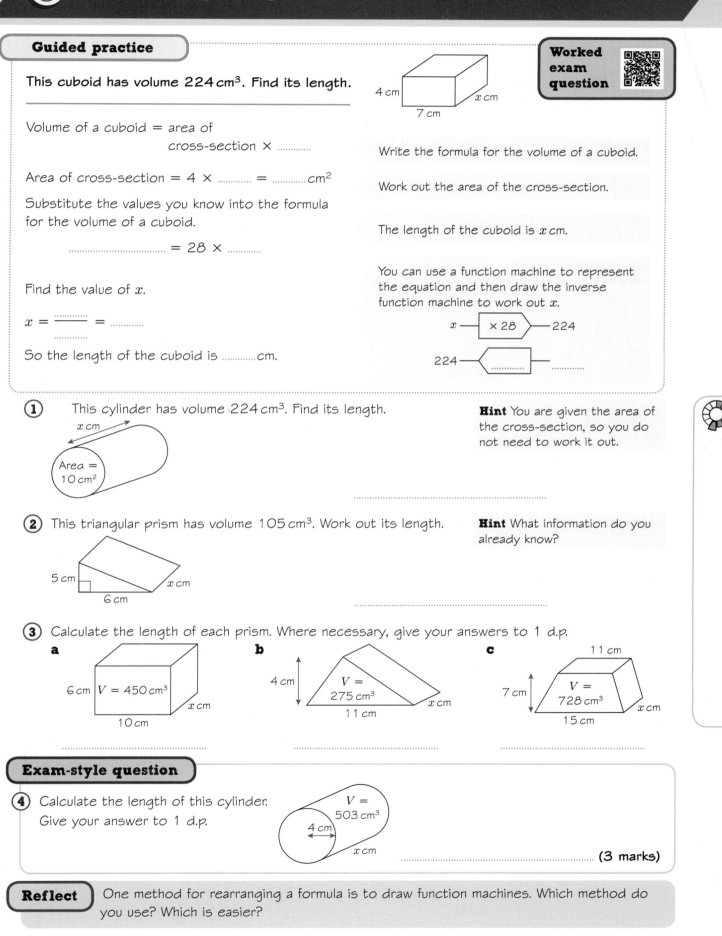

Guided practice

This cuboid has volume 224 cm³. Find its length.

Volume of a cuboid = area of
cross-section × ....

Area of cross-section = 4 × .... = ... cm²

Substitute the values you know into the formula for the volume of a cuboid.

.... = 28 × ....

Find the value of x.

$x = \dfrac{............}{............} = ...$

So the length of the cuboid is .... cm.

Worked exam question

Write the formula for the volume of a cuboid.

Work out the area of the cross-section.

The length of the cuboid is x cm.

You can use a function machine to represent the equation and then draw the inverse function machine to work out x.

$x \longrightarrow \boxed{\times 28} \longrightarrow 224$

$224 \longrightarrow \boxed{} \longrightarrow ....$

---

① This cylinder has volume 224 cm³. Find its length.

Area = 10 cm²

Hint You are given the area of the cross-section, so you do not need to work it out.

---

② This triangular prism has volume 105 cm³. Work out its length.

5 cm
6 cm
x cm

Hint What information do you already know?

---

③ Calculate the length of each prism. Where necessary, give your answers to 1 d.p.

a
6 cm V = 450 cm³
10 cm
x cm

b
4 cm
V = 275 cm³
11 cm
x cm

c
11 cm
7 cm V = 728 cm³
15 cm
x cm

Exam-style question

④ Calculate the length of this cylinder.
Give your answer to 1 d.p.

V = 503 cm³
4 cm
x cm

.... (3 marks)

Reflect One method for rearranging a formula is to draw function machines. Which method do you use? Which is easier?

Practise the methods

Answer this question to check where to start.

Check up

Tick the correct combination of faces for this solid.

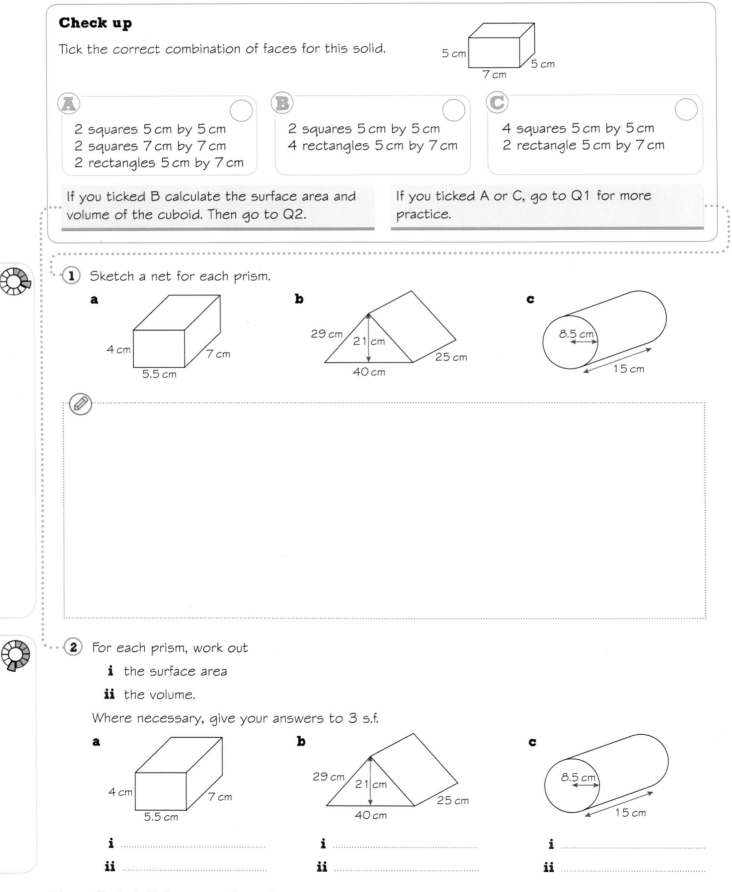

5 cm
7 cm
5 cm

A ◯

2 squares 5 cm by 5 cm
2 squares 7 cm by 7 cm
2 rectangles 5 cm by 7 cm

B ◯

2 squares 5 cm by 5 cm
4 rectangles 5 cm by 7 cm

C ◯

4 squares 5 cm by 5 cm
2 rectangle 5 cm by 7 cm

If you ticked B calculate the surface area and volume of the cuboid. Then go to Q2.

If you ticked A or C, go to Q1 for more practice.

① Sketch a net for each prism.

a
4 cm
7 cm
5.5 cm

b
29 cm
21 cm
40 cm
25 cm

c
8.5 cm
15 cm

② For each prism, work out

 i the surface area

 ii the volume.

Where necessary, give your answers to 3 s.f.

a
4 cm
7 cm
5.5 cm

b
29 cm
21 cm
40 cm
25 cm

c
8.5 cm
15 cm

a
 i ....
 ii ....

b
 i ....
 ii ....

c
 i ....
 ii ....

(3) For each solid, work out

    i the surface area

    ii the volume.

Where necessary, give your answers to 3 s.f.

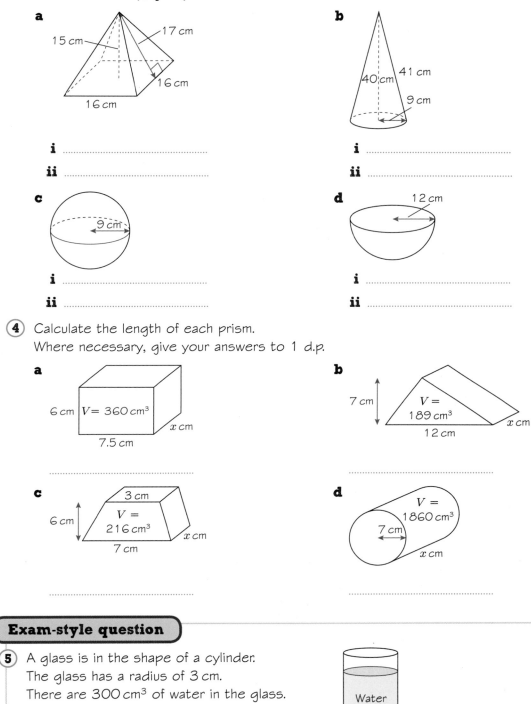

a

15 cm — 17 cm

16 cm

16 cm

i ....

ii ....

b

40 cm — 41 cm

9 cm

i ....

ii ....

c

9 cm

i ....

ii ....

d

12 cm

i ....

ii ....

(4) Calculate the length of each prism.
Where necessary, give your answers to 1 d.p.

a

6 cm | $V = 360\,cm^3$

7.5 cm

x cm

....

b

7 cm

$V = 189\,cm^3$

12 cm

x cm

....

c

3 cm

6 cm

$V = 216\,cm^3$

7 cm

x cm

....

d

$V = 1860\,cm^3$

7 cm

x cm

....

(5) A glass is in the shape of a cylinder.
The glass has a radius of 3 cm.
There are $300\,cm^3$ of water in the glass.
Work out the depth of the water in the glass.
Give your answer to 1 d.p.

Water

3 cm

.... **(3 marks)**

Problem-solve!

① A matchbox is 4 cm by 6 cm by 1.5 cm.

A carton is 20 cm by 30 cm by 12 cm.

The carton is completely filled with matchboxes.

Work out the number of matchboxes in the carton.

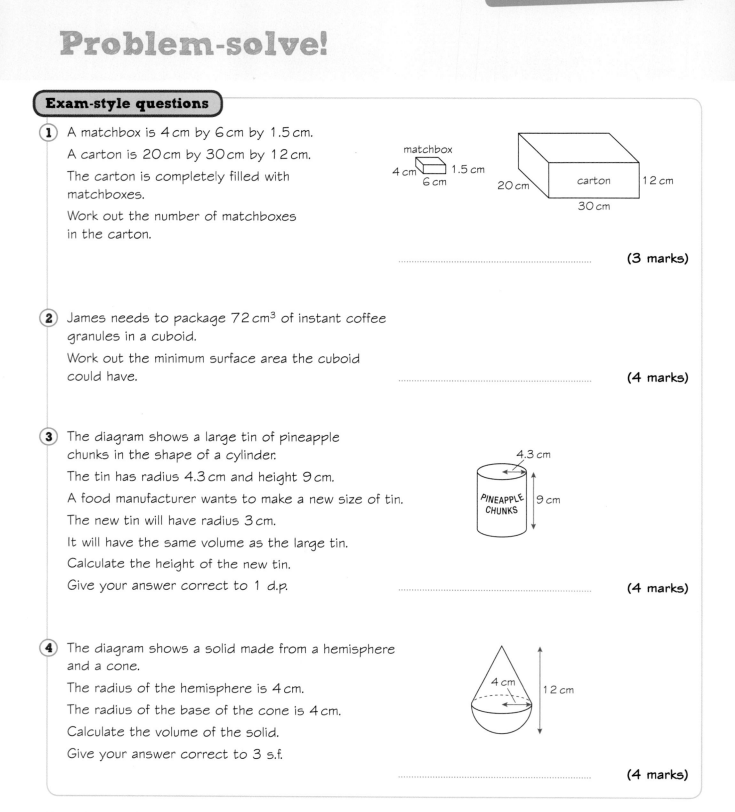

matchbox

4 cm   1.5 cm
   6 cm

20 cm   carton   12 cm
      30 cm

.... (3 marks)

② James needs to package 72 cm³ of instant coffee granules in a cuboid.

Work out the minimum surface area the cuboid could have.

.... (4 marks)

③ The diagram shows a large tin of pineapple chunks in the shape of a cylinder.

The tin has radius 4.3 cm and height 9 cm.

A food manufacturer wants to make a new size of tin.

The new tin will have radius 3 cm.

It will have the same volume as the large tin.

Calculate the height of the new tin.

Give your answer correct to 1 d.p.

4.3 cm

PINEAPPLE
CHUNKS   9 cm

.... (4 marks)

④ The diagram shows a solid made from a hemisphere and a cone.

The radius of the hemisphere is 4 cm.

The radius of the base of the cone is 4 cm.

Calculate the volume of the solid.

Give your answer correct to 3 s.f.

4 cm   12 cm

.... (4 marks)

Now that you have completed this unit, how confident do you feel?

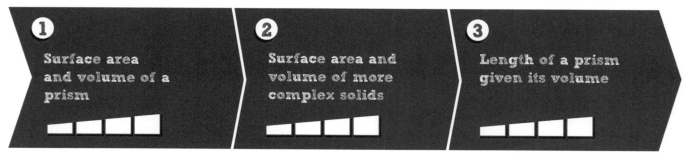

① Surface area and volume of a prism

② Surface area and volume of more complex solids

③ Length of a prism given its volume

3 Angles

This unit will help you work out interior and exterior angles and use the angle sum of a polygon.

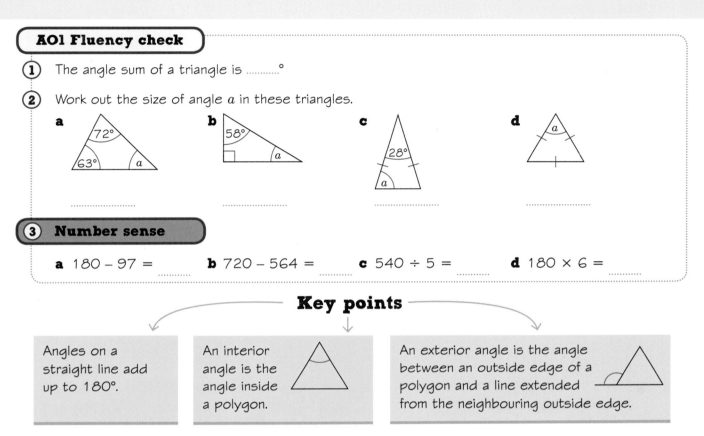

AO1 Fluency check

(1) The angle sum of a triangle is ....°

(2) Work out the size of angle a in these triangles.

a 72° 63° a **b** 58° a **c** 28° a **d** a

(3) **Number sense**

  a $180 - 97 =$ **b** $720 - 564 =$ **c** $540 \div 5 =$ **d** $180 \times 6 =$

Key points

Angles on a straight line add up to 180°.

An interior angle is the angle inside a polygon.

An exterior angle is the angle between an outside edge of a polygon and a line extended from the neighbouring outside edge.

These **skills boosts** will help you solve problems involving angles.

1 Interior and exterior angles   2 Angle sum of a polygon   3 Using the angle sum of a polygon

You might have already done some work using angles. Before starting the first skills boost, rate your confidence working with different types of angles.

1 What is the exterior angle of an equilateral triangle?

2 What is the sum of the interior angles in a quadrilateral?

3 Work out the interior angle of a regular pentagon.

How confident are you?

1 Interior and exterior angles

The exterior angles of a polygon always add up to 360°.

Guided practice

Work out the size of angle p.

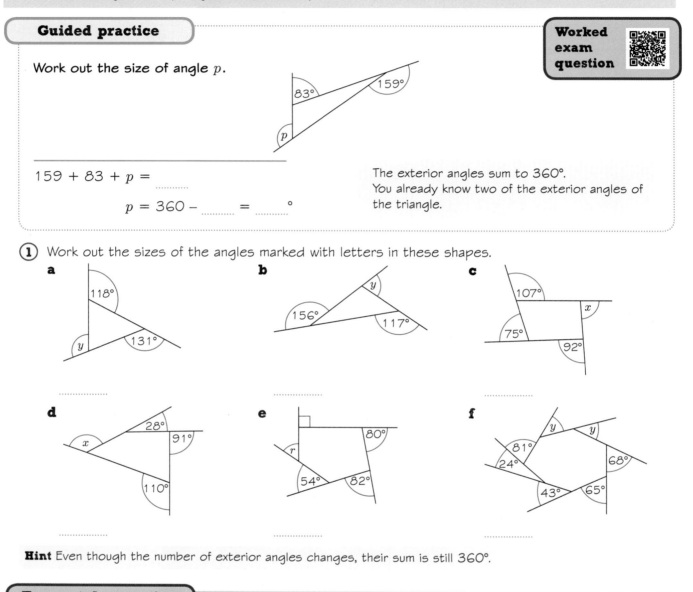

83°   159°

p

$159 + 83 + p =$ ....

$p = 360 -$ .... $=$ .... °

The exterior angles sum to 360°.
You already know two of the exterior angles of the triangle.

Worked exam question

① Work out the sizes of the angles marked with letters in these shapes.

a

118°

y 131°

....

b

y

156°   117°

....

c

107°

x

75°

92°

....

d

28°   91°

x

110°

....

e

r 80°

54°   82°

....

f

y y

81°   68°

24°

43°   65°

....

Hint Even though the number of exterior angles changes, their sum is still 360°.

Exam-style question

② The diagram shows an exterior and an interior angle of a regular polygon.
Work out

exterior angle   interior angle

45°

a the number of sides of the polygon   .... (2 marks)

b the size of the interior angle.   .... (1 mark)

③ The exterior angle of a regular polygon is 72°.
Work out

a the number of sides of the polygon   ....

b the size of the interior angle.   ....

Hint Draw a sketch of the exterior and interior angles similar to the diagram in Q2.

Reflect What do an interior angle and its exterior angle sum to?

2 Angle sum of a polygon

To find the sum of the interior angles of a polygon, you can divide the polygon into triangles. The angles in each triangle sum to 180°.

Guided practice

Work out the sum of the interior angles in a pentagon.

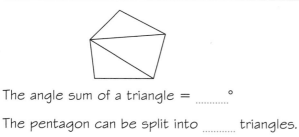

Using just one corner, split the pentagon completely into triangles. Multiply the number of triangles by the angle sum of one triangle.

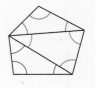

The angle sum of a triangle = ... °

The pentagon can be split into .... triangles.

Angle sum of a pentagon = ... × .... = ... °

(1) Work out the sum of the interior angles in a quadrilateral. ....

Hint How many triangles can you divide the quadrilateral into?

(2) **a** Complete the table to work out the angle sum for each polygon.

Number of sides	Number of triangles	Interior angle sum
3	1	1 × 180° = 180°
4	2	2 × 180° = 360°
5		
6		
7		
8	6	

Hint Split the shapes into triangles.

b Complete the rule for working out the angle sum.

Interior angle sum of a polygon = (number of sides − ....) × .... °

Hint Look for the pattern in the table.

Exam-style question

(3) Work out the interior angle sum of a polygon with

a 16 sides .... (2 marks)

b 20 sides. .... (2 marks)

(4) How many sides does a polygon have when its interior angles sum to
a 1260°   **b** 2340°?

Hint Look at the rule you worked out in Q2b.

....

Reflect Without looking at your answer to Q2b, write the rule for working out the interior angle sum of any polygon.

3 Using the angle sum of a polygon

The interior angles of an n-sided polygon sum to $(n - 2) \times 180°$.

Guided practice

Work out the size of the interior angle of a regular hexagon.

Angle sum $= (n - 2) \times 180°$

A hexagon has 6 sides so put n equal to 6.

Angle sum $= \underline{\hspace{1cm}} \times 180° = \underline{\hspace{1cm}}°$

Each interior angle $= \dfrac{\overline{\hspace{1cm}}}{6} = \underline{\hspace{1cm}}°$

A hexagon has 6 angles so divide the sum of all the interior angles by 6.

1. Work out the size of the interior angle of a regular pentagon.

   ....

2. Work out the size of the interior angle of

   a a regular octagon   **b** a regular nonagon   **c** a regular 16-sided polygon.

   ....

3. Work out the sizes of the angles marked with letters.

   Hint First work out the sum of the interior angles in the polygon.

   a
   97°
   102°
   95°
   a

   b
   b
   88°
   136°
   153°
   102° 115°

   c
   145°
   143°
   119°
   128°
   137°
   128°
   c   c

   ....

Exam-style question

4. Work out the size of the angles marked with the letter d.

   d
   d
   d
   d

   .... **(3 marks)**

Reflect

Explain the difference between the angles in a regular polygon and the angles in an irregular polygon.

Practise the methods

Answer this question to check where to start.

Check up

Tick the triangle with its exterior angles correctly identified.

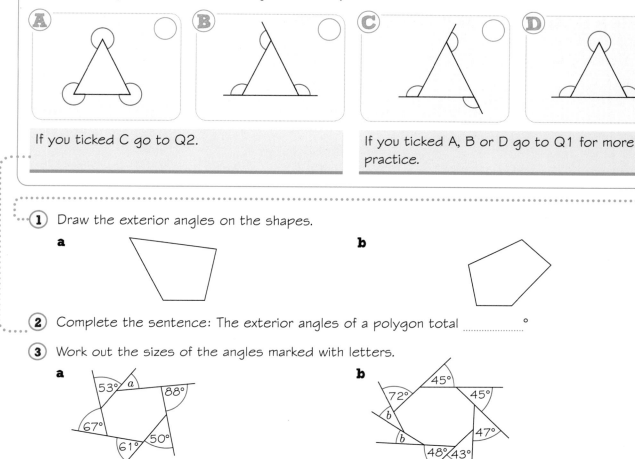

A ○ **B** ○ **C** ○ **D** ○

| If you ticked C go to Q2. | If you ticked A, B or D go to Q1 for more practice. |

1 Draw the exterior angles on the shapes.

a

b

2 Complete the sentence: The exterior angles of a polygon total .... °

3 Work out the sizes of the angles marked with letters.

a
53° a 88°
67°
61° 50°

....

b
45°
72° 45°
b
b
47°
48° 43°

....

4 The exterior angle of a regular polygon is 40°. Work out

a the number of sides of the polygon   **b** the size of the interior angle.

....

5 Work out the sum of the interior angles of a polygon with 18 sides. ....

Exam-style question

6 Work out the size of the interior angles of a regular decagon. .... **(3 marks)**

7 Work out the size of angle a.

151° 72°
163°
108° a
116° 165°

....

Problem-solve!

(1) The interior angles of a polygon sum to 3060°.

How many sides does it have? ....

Exam-style questions

(2) The diagram shows an equilateral triangle and 3 regular pentagons.
Work out the size of the angle marked x.

.... (3 marks)

(3) ABCDEFGHIJ is a regular decagon.
AFKL is a square.
Work out the size of angle KFG.

.... (4 marks)

(4) A, B and C are three vertices of a regular octagon.
Angle ABX = 90°.
Work out the size of angle CBX.

.... (3 marks)

(5) The diagram shows two regular polygons.
Find the size of the angle marked x.
Give reasons for your answer.

.... (4 marks)

(6) ABCDEFGH is a regular octagon.
ABKEFJ is a hexagon.
JK is a line of symmetry of the hexagon.
Angle AJF = angle BKE = 160°.
Work out the size of angle KED.
Give reasons for your answer.

.... (4 marks)

(7) The interior angle of a regular polygon is 160°.
Work out the number of sides of the polygon.

.... (3 marks)

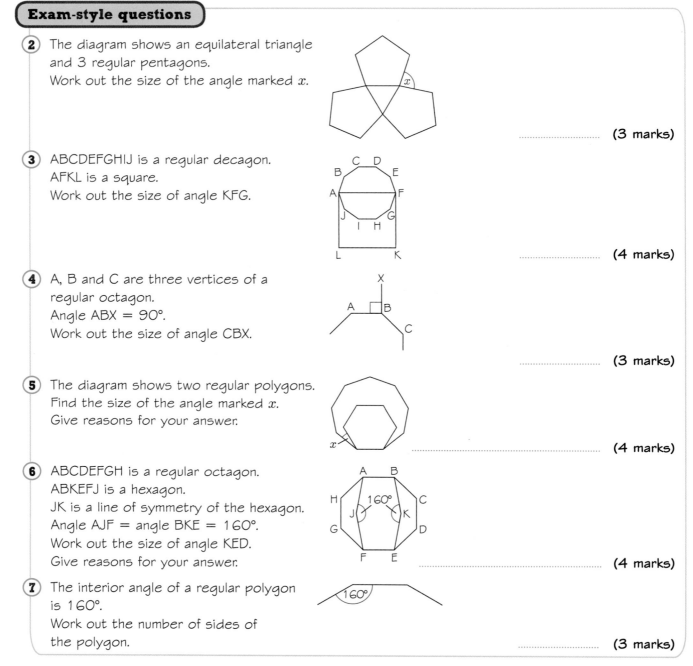

Now that you have completed this unit, how confident do you feel?

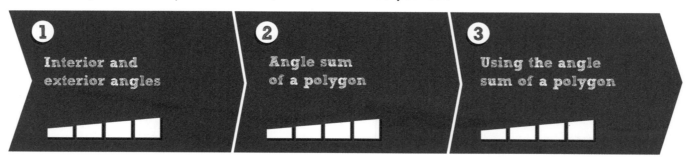

1 Interior and exterior angles

2 Angle sum of a polygon

3 Using the angle sum of a polygon

④ Vectors

This unit will help you understand and use vector notation and calculate using vectors.

AO1 Fluency check

① Simplify

  a $3a + 5a$　　　　**b** $-b + 3b$　　　　**c** $4p + 2q - p + 3q$　　**d** $-4x + 5y - 6x$

  ....　　........................　　........................　　........................

② **Number sense**

Work out

  a $-4 + 7$ ....　**b** $-3 - 6$ ....　**c** $-2 - -8$ ....　**d** $-1 + -4$ ....

  e -3×5 ....　**f** 8×-9 ....　**g** -5×-12 ....　**h** -4×6 ....

— Key points —

A vector describes the distance and direction from one point to another.	There are different ways of using notation to describe vectors.

These **skills boosts** will help you solve problems using vectors.

> ❶ **Understand and use vector notation**　❷ **Add and subtract vectors**　❸ **Find multiples of vectors**

You might have already done some work using vectors. Before starting the first skills boost, rate your confidence using vectors in different ways.

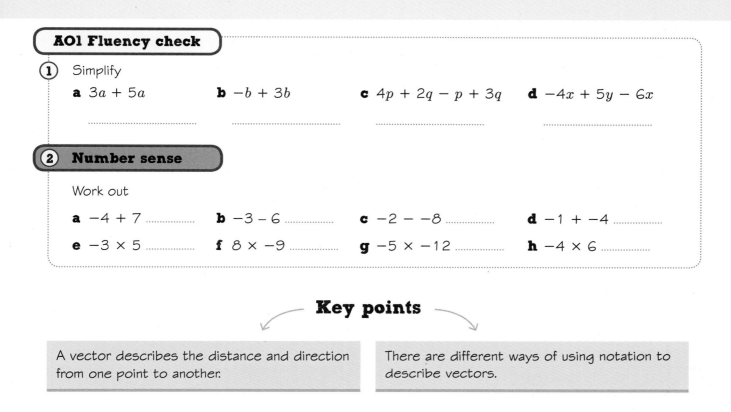

① Write $\overrightarrow{XY}$ as a column vector.

② Work out $\begin{pmatrix} 3 \\ 2 \end{pmatrix} + \begin{pmatrix} 1 \\ 4 \end{pmatrix}$.

③ The vector **b** is $\begin{pmatrix} 2 \\ 5 \end{pmatrix}$.

Write 4**b** as a column vector.

How confident are you?

1 Understand and use vector notation

A vector can be written as a column vector, for example $\begin{pmatrix} 3 \\ 2 \end{pmatrix}$.

A vector from A to B can be written as $\overrightarrow{AB}$.

A vector can also be written as a bold lower-case letter, for example **a**. When handwriting a vector using lower-case letters, underline the letter, for example <u>a</u>.

Guided practice

Write these vectors as column vectors.

a $\overrightarrow{AB}$

b $\overrightarrow{CD}$

c $\overrightarrow{EF}$

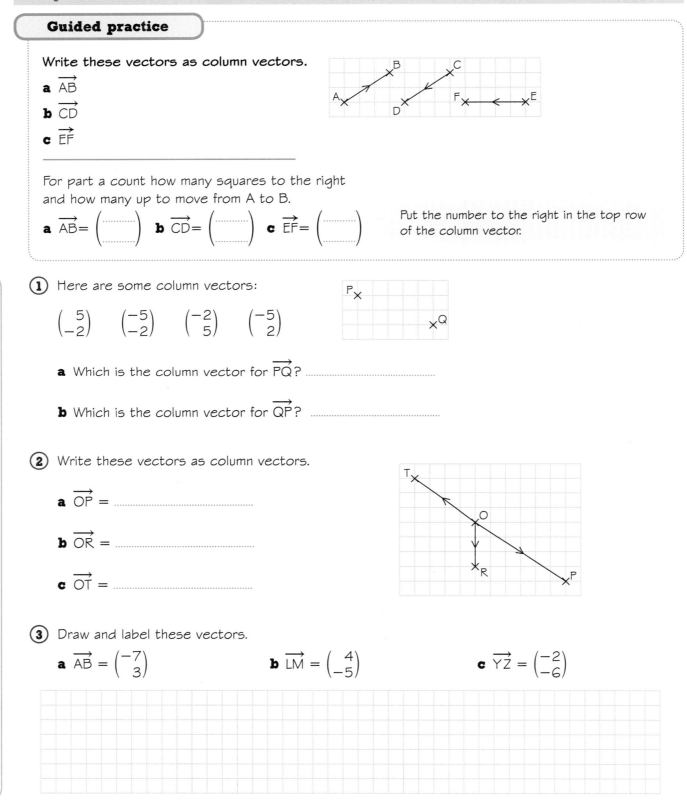

For part a count how many squares to the right and how many up to move from A to B.

a $\overrightarrow{AB} = \begin{pmatrix} \dots \\ \dots \end{pmatrix}$ **b** $\overrightarrow{CD} = \begin{pmatrix} \dots \\ \dots \end{pmatrix}$ **c** $\overrightarrow{EF} = \begin{pmatrix} \dots \\ \dots \end{pmatrix}$

Put the number to the right in the top row of the column vector.

(1) Here are some column vectors:

$\begin{pmatrix} 5 \\ -2 \end{pmatrix}$ $\begin{pmatrix} -5 \\ -2 \end{pmatrix}$ $\begin{pmatrix} -2 \\ 5 \end{pmatrix}$ $\begin{pmatrix} -5 \\ 2 \end{pmatrix}$

a Which is the column vector for $\overrightarrow{PQ}$? ....

b Which is the column vector for $\overrightarrow{QP}$? ....

(2) Write these vectors as column vectors.

a $\overrightarrow{OP}$ = ...

b $\overrightarrow{OR}$ = ...

c $\overrightarrow{OT}$ = ...

(3) Draw and label these vectors.

a $\overrightarrow{AB} = \begin{pmatrix} -7 \\ 3 \end{pmatrix}$ **b** $\overrightarrow{LM} = \begin{pmatrix} 4 \\ -5 \end{pmatrix}$ **c** $\overrightarrow{YZ} = \begin{pmatrix} -2 \\ -6 \end{pmatrix}$

④ Draw and label these vectors.

a $a = \begin{pmatrix} 5 \\ 1 \end{pmatrix}$ **b** $b = \begin{pmatrix} -3 \\ 3 \end{pmatrix}$

c $c = \begin{pmatrix} 0 \\ -5 \end{pmatrix}$ **d** $d = \begin{pmatrix} -2 \\ -2 \end{pmatrix}$

⑤ Write these vectors as column vectors.

a a ....

b −a ....

c What do you notice about your answers to parts a and b?

....

⑥ $a = \begin{pmatrix} 2 \\ 3 \end{pmatrix}$ $b = \begin{pmatrix} 3 \\ 2 \end{pmatrix}$ $c = \begin{pmatrix} 3 \\ -2 \end{pmatrix}$

Label these vectors using **a, b, c, −a, −b, −c**.

Hint −**a** is the negative of **a** and points in the opposite direction.

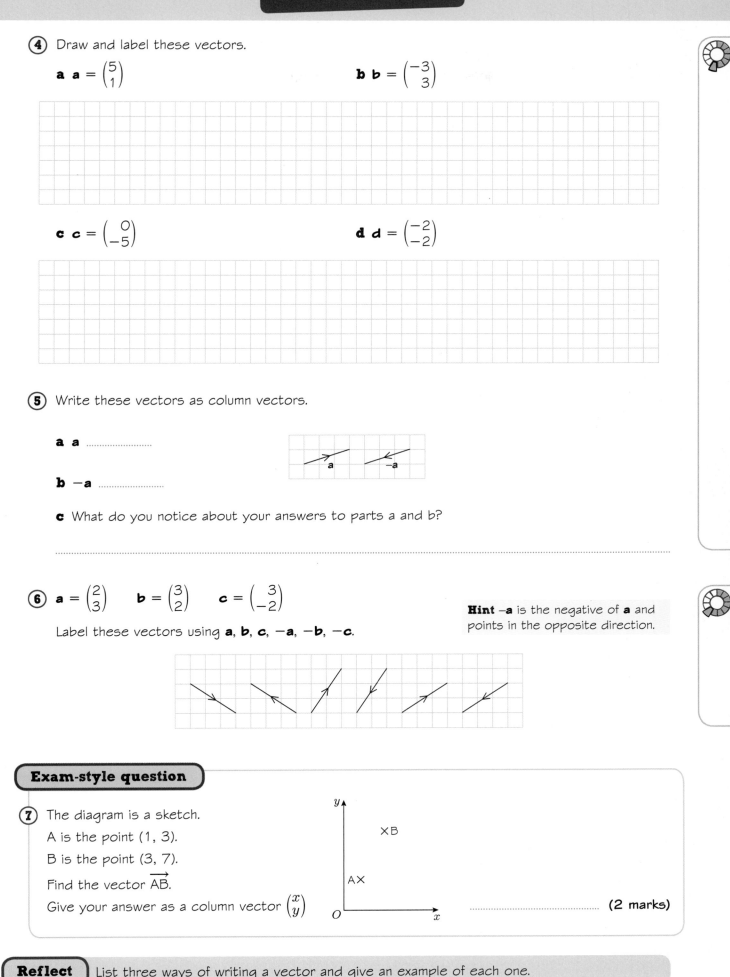

Exam-style question

⑦ The diagram is a sketch.

A is the point (1, 3).

B is the point (3, 7).

Find the vector $\overrightarrow{AB}$.

Give your answer as a column vector $\begin{pmatrix} x \\ y \end{pmatrix}$

.... (2 marks)

Reflect   List three ways of writing a vector and give an example of each one.

2 Add and subtract vectors

Guided practice

Add these column vectors. Give your answer as a single column vector.

$$\begin{pmatrix} 4 \\ 1 \end{pmatrix} + \begin{pmatrix} -1 \\ 3 \end{pmatrix}$$

$$\begin{pmatrix} 4 \\ 1 \end{pmatrix} + \begin{pmatrix} -1 \\ 3 \end{pmatrix} = \begin{pmatrix} 4 + -1 \\ 1 + 3 \end{pmatrix} = \begin{pmatrix} 3 \\ \ldots \end{pmatrix}$$

Add the top numbers and add the bottom numbers.

(1) Add these column vectors. Give your answer as a single column vector.

a $\begin{pmatrix} -2 \\ 0 \end{pmatrix} + \begin{pmatrix} -3 \\ -7 \end{pmatrix} =$ ....

b $\begin{pmatrix} 3 \\ 5 \end{pmatrix} + \begin{pmatrix} -8 \\ -1 \end{pmatrix} =$ ....

c $\begin{pmatrix} -3 \\ 2 \end{pmatrix} + \begin{pmatrix} -6 \\ 5 \end{pmatrix} =$ ....

(2) Use the diagram to write these vectors as column vectors.

a x = ...

b y = ...

c z = ...

d i Use your column vectors from parts a and b to work out **x + y** ....

ii Use your answer to show that **x + y = z** ....

Exam-style question

(3) $\mathbf{a} = \begin{pmatrix} 2 \\ -3 \end{pmatrix}$, $\mathbf{b} = \begin{pmatrix} -1 \\ -3 \end{pmatrix}$ and $\mathbf{c} = \mathbf{a} + \mathbf{b}$.

a Write **c** as a column vector. .... (2 marks)

b Draw a diagram to show **a**, **b** and **c** as a triangle.

(3 marks)

(4) Subtract these column vectors. Give your answer as a single column vector.

a $\begin{pmatrix} 4 \\ 7 \end{pmatrix} - \begin{pmatrix} 2 \\ 8 \end{pmatrix} =$ ....

b $\begin{pmatrix} 1 \\ 3 \end{pmatrix} - \begin{pmatrix} 6 \\ -1 \end{pmatrix} =$ ....

c $\begin{pmatrix} -2 \\ -5 \end{pmatrix} - \begin{pmatrix} -3 \\ 2 \end{pmatrix} =$ ....

Hint Subtract the top numbers and subtract the bottom numbers.

(5) $\mathbf{p} = \begin{pmatrix} 2 \\ 1 \end{pmatrix}$, $\mathbf{q} = \begin{pmatrix} 4 \\ 4 \end{pmatrix}$ and $\mathbf{r} = \mathbf{p} - \mathbf{q}$.

a Write **r** as a column vector. ....

b Draw a diagram to show **p**, **−q** and **r** as a triangle.

Hint −**q** is the negative of **q** and so points in the opposite direction.

Reflect Write a rule for adding and subtracting column vectors.

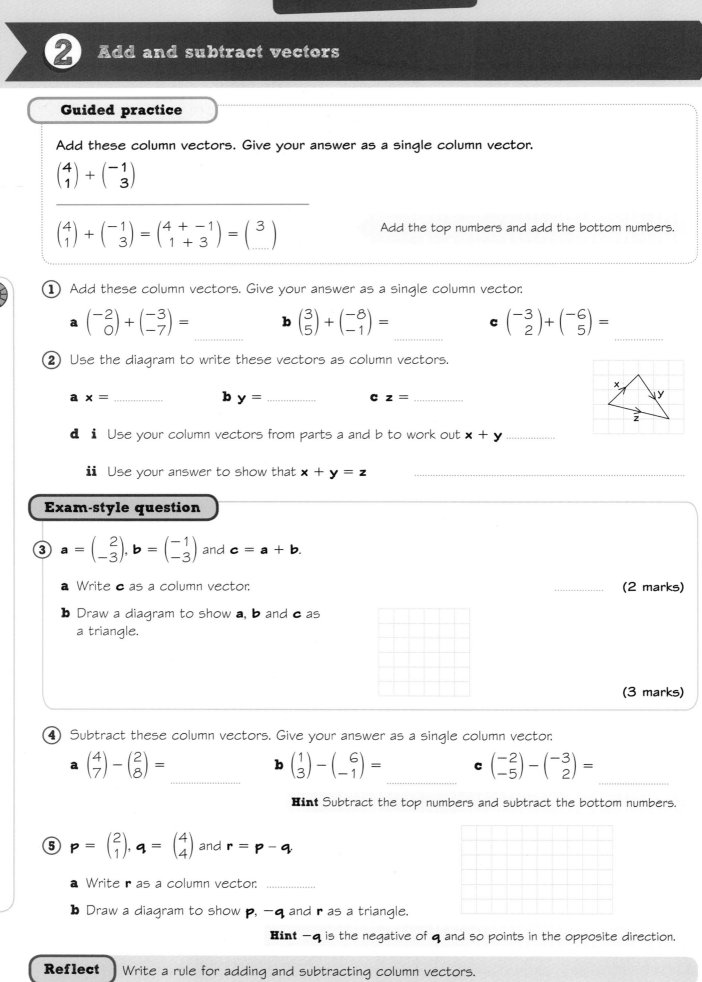

3 Find multiples of vectors

Multiples of vectors can be found by multiplying the top and the bottom numbers of the column vector by the same number.

Guided practice

Write the column vector

a **a**

b 2**a**.

_____

a $a = \begin{pmatrix} \text{.......} \\ \text{.......} \end{pmatrix}$ **a** is 3 squares right and 2 squares down.

b $2a = 2 \times \begin{pmatrix} \text{.......} \\ \text{.......} \end{pmatrix}$

   Multiply each number in **a** by 2.

   $= \begin{pmatrix} 2 \times 3 \\ 2 \times \text{.....} \end{pmatrix}$ **2a** is twice the distance right and twice the distance down.

   $= \begin{pmatrix} \text{.......} \\ \text{.......} \end{pmatrix}$

① $x = \begin{pmatrix} 0 \\ -4 \end{pmatrix}$ $y = \begin{pmatrix} -2 \\ 5 \end{pmatrix}$

Write as column vectors

a $-2x = $ ....   **b** $3y = $ ....   **c** $3x = $ ....   **d** $-y = $ ....

② $p = \begin{pmatrix} -1 \\ 2 \end{pmatrix}$ $q = \begin{pmatrix} 2 \\ -3 \end{pmatrix}$

Write as column vectors

a $2p = $ ....   **b** $-3q = $ ....   **c** $2p + q = $ ....   **d** $3q - p = $ ....

Exam-style question

③ $r = \begin{pmatrix} -6 \\ 2 \end{pmatrix}$

Write as column vectors

a a vector in the same direction as **r** but half as long   .... **(1 mark)**

b a vector the same length as **r** but in the opposite direction   .... **(1 mark)**

c a vector half as long as **r** and in the opposite direction.   .... **(1 mark)**

Reflect   If two vectors are parallel but point in opposite directions and one is 3 times as long as the other, what is the relationship between the column vectors?

Practise the methods

Answer this question to check where to start.

Check up

Tick the correct column vector for $-2p$ when $p = \begin{pmatrix} -3 \\ 1 \end{pmatrix}$

A $\begin{pmatrix} 6 \\ -1 \end{pmatrix}$ ○   **B** $\begin{pmatrix} 6 \\ -2 \end{pmatrix}$ ○   **C** $\begin{pmatrix} -6 \\ 2 \end{pmatrix}$ ○   **D** $\begin{pmatrix} -6 \\ -2 \end{pmatrix}$ ○

If you ticked B then go to Q3.   If you ticked A, C or D go to Q1 for more practice.

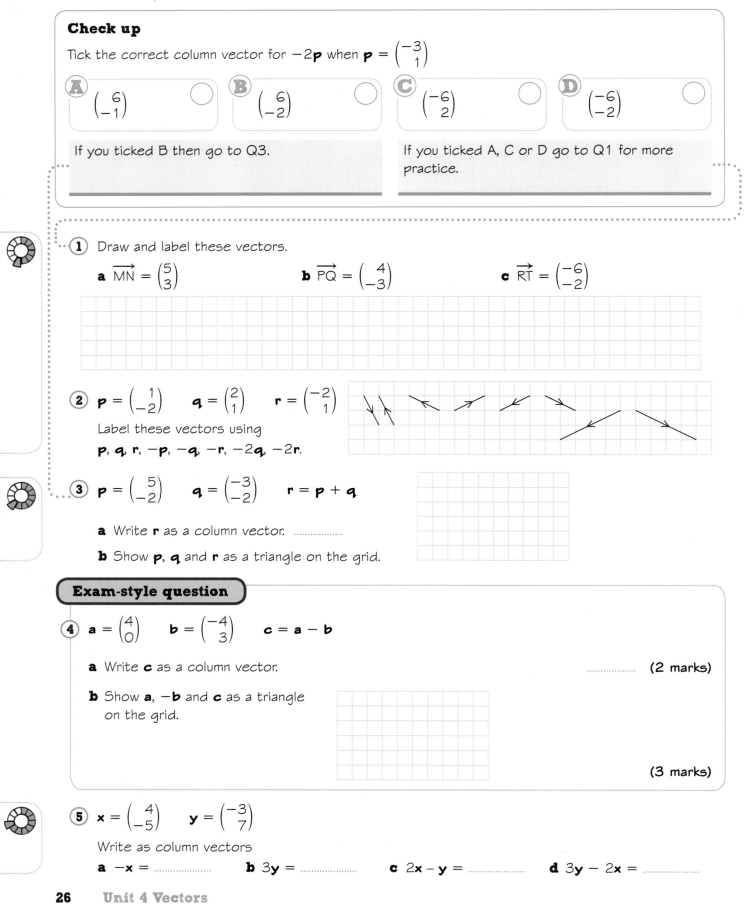

① Draw and label these vectors.

a $\overrightarrow{MN} = \begin{pmatrix} 5 \\ 3 \end{pmatrix}$ **b** $\overrightarrow{PQ} = \begin{pmatrix} 4 \\ -3 \end{pmatrix}$ **c** $\overrightarrow{RT} = \begin{pmatrix} -6 \\ -2 \end{pmatrix}$

② $p = \begin{pmatrix} 1 \\ -2 \end{pmatrix}$ $q = \begin{pmatrix} 2 \\ 1 \end{pmatrix}$ $r = \begin{pmatrix} -2 \\ 1 \end{pmatrix}$

Label these vectors using

$p, q, r, -p, -q, -r, -2q, -2r$.

③ $p = \begin{pmatrix} 5 \\ -2 \end{pmatrix}$ $q = \begin{pmatrix} -3 \\ -2 \end{pmatrix}$ $r = p + q$

a Write r as a column vector. ....

b Show p, q and r as a triangle on the grid.

Exam-style question

④ $a = \begin{pmatrix} 4 \\ 0 \end{pmatrix}$ $b = \begin{pmatrix} -4 \\ 3 \end{pmatrix}$ $c = a - b$

a Write c as a column vector.   .... (2 marks)

b Show $a, -b$ and c as a triangle on the grid.

(3 marks)

⑤ $x = \begin{pmatrix} 4 \\ -5 \end{pmatrix}$ $y = \begin{pmatrix} -3 \\ 7 \end{pmatrix}$

Write as column vectors

a $-x =$ ....   **b** $3y =$ ....   **c** $2x - y =$ ....   **d** $3y - 2x =$ ....

Problem-solve!

(1)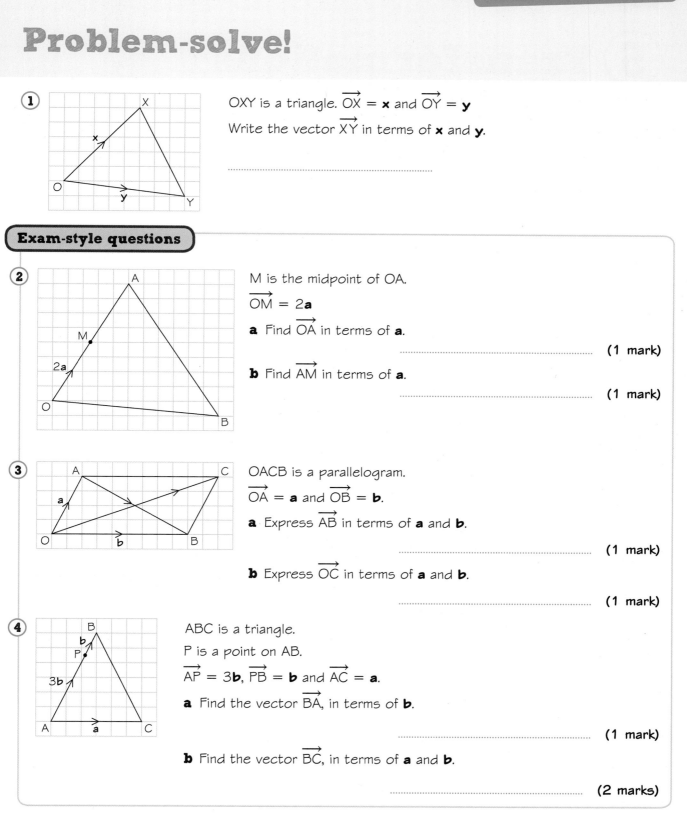

OXY is a triangle. $\overrightarrow{OX} = $ **x** and $\overrightarrow{OY} = $ **y**

Write the vector $\overrightarrow{XY}$ in terms of **x** and **y**.

....

Exam-style questions

(2)

M is the midpoint of OA.

$\overrightarrow{OM} = 2$**a**

a Find $\overrightarrow{OA}$ in terms of **a**.

.... (1 mark)

b Find $\overrightarrow{AM}$ in terms of **a**.

.... (1 mark)

(3)

OACB is a parallelogram.

$\overrightarrow{OA} = $ **a** and $\overrightarrow{OB} = $ **b**.

a Express $\overrightarrow{AB}$ in terms of **a** and **b**.

.... (1 mark)

b Express $\overrightarrow{OC}$ in terms of **a** and **b**.

.... (1 mark)

(4)

ABC is a triangle.

P is a point on AB.

$\overrightarrow{AP} = 3$**b**, $\overrightarrow{PB} = $ **b** and $\overrightarrow{AC} = $ **a**.

a Find the vector $\overrightarrow{BA}$, in terms of **b**.

.... (1 mark)

b Find the vector $\overrightarrow{BC}$, in terms of **a** and **b**.

.... (2 marks)

Now that you have completed this unit, how confident do you feel?

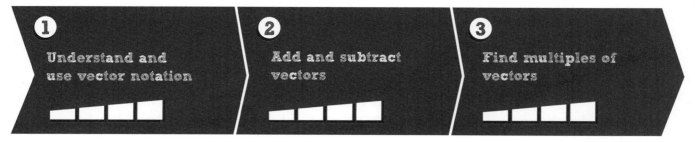

1 Understand and use vector notation

2 Add and subtract vectors

3 Find multiples of vectors

⑤ Transformations

This unit will help you translate and enlarge shapes and describe translations and enlargements.

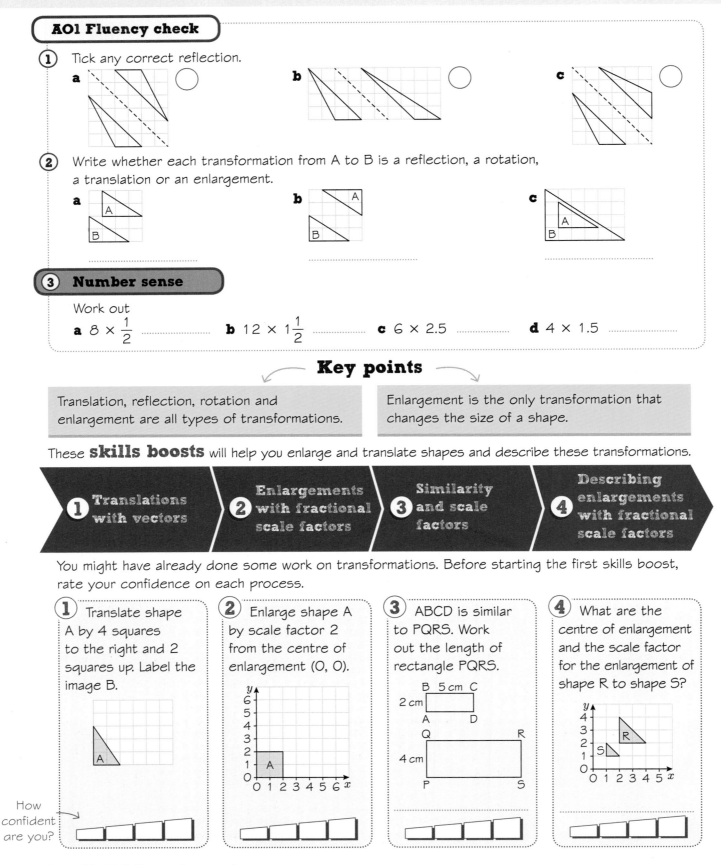

AO1 Fluency check

① Tick any correct reflection.

a ⬭   b ⬭   c ⬭

② Write whether each transformation from A to B is a reflection, a rotation, a translation or an enlargement.

a   b   c

③ **Number sense**

Work out

a $8 \times \frac{1}{2}$ ....   b $12 \times 1\frac{1}{2}$ ....   c 6×2.5 ....   d 4×1.5 ....

Key points

Translation, reflection, rotation and enlargement are all types of transformations.

Enlargement is the only transformation that changes the size of a shape.

These **skills boosts** will help you enlarge and translate shapes and describe these transformations.

① **Translations with vectors**   ② **Enlargements with fractional scale factors**   ③ **Similarity and scale factors**   ④ **Describing enlargements with fractional scale factors**

You might have already done some work on transformations. Before starting the first skills boost, rate your confidence on each process.

① Translate shape A by 4 squares to the right and 2 squares up. Label the image B.

② Enlarge shape A by scale factor 2 from the centre of enlargement (0, 0).

③ ABCD is similar to PQRS. Work out the length of rectangle PQRS.

④ What are the centre of enlargement and the scale factor for the enlargement of shape R to shape S?

How confident are you?

1 Translations with vectors

A translation can be described using a column vector. A translation by $\binom{a}{b}$ means a squares to the right and b squares up.

Guided practice

Translate shape A by $\binom{3}{1}$.

Label the image X.

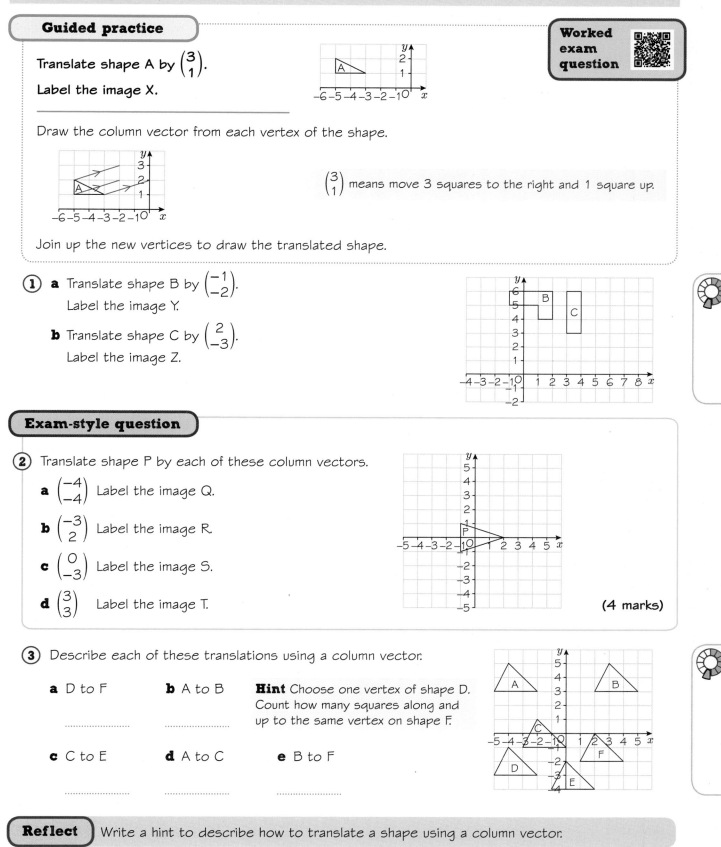

Draw the column vector from each vertex of the shape.

$\binom{3}{1}$ means move 3 squares to the right and 1 square up.

Join up the new vertices to draw the translated shape.

Worked exam question

① **a** Translate shape B by $\binom{-1}{-2}$.

Label the image Y.

b Translate shape C by $\binom{2}{-3}$.

Label the image Z.

Exam-style question

② Translate shape P by each of these column vectors.

a $\binom{-4}{-4}$ Label the image Q.

b $\binom{-3}{2}$ Label the image R.

c $\binom{0}{-3}$ Label the image S.

d $\binom{3}{3}$ Label the image T.

(4 marks)

③ Describe each of these translations using a column vector.

a D to F   **b** A to B   **Hint** Choose one vertex of shape D.
Count how many squares along and
up to the same vertex on shape F.

....

c C to E   **d** A to C   **e** B to F

....

Reflect

Write a hint to describe how to translate a shape using a column vector.

2 Enlargements with fractional scale factors

Shapes can be enlarged by fractional scale factors. If the scale factor is less than 1, then the enlarged shape will be smaller than the original shape.

Guided practice

Enlarge shape A by scale factor $\frac{1}{2}$ with **centre of enlargement** (0, 0).

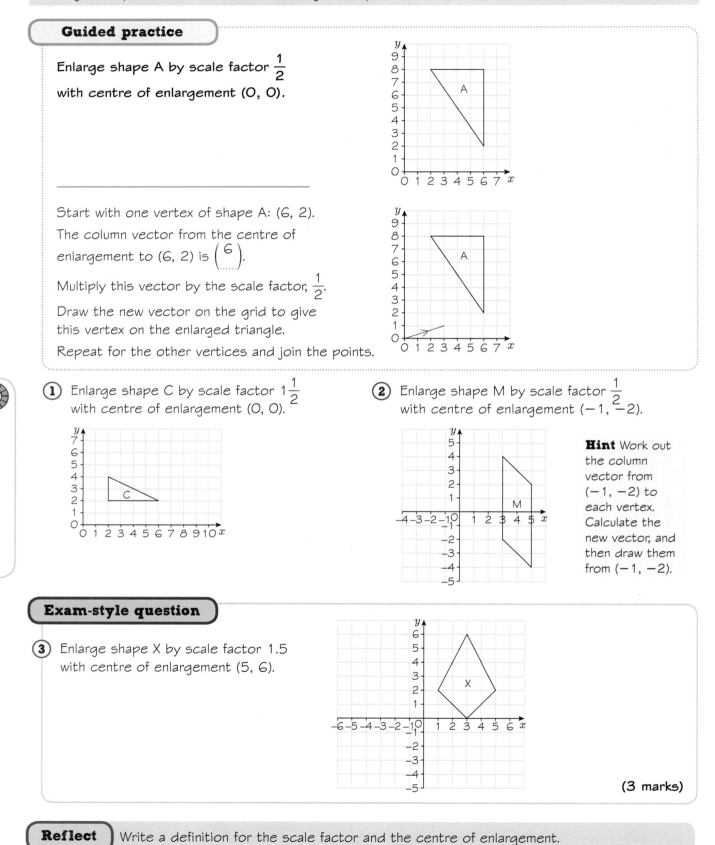

Start with one vertex of shape A: (6, 2).

The column vector from the centre of enlargement to (6, 2) is $\begin{pmatrix} 6 \\ \end{pmatrix}$.

Multiply this vector by the scale factor, $\frac{1}{2}$.

Draw the new vector on the grid to give this vertex on the enlarged triangle.

Repeat for the other vertices and join the points.

① Enlarge shape C by scale factor $1\frac{1}{2}$ with centre of enlargement (0, 0).

② Enlarge shape M by scale factor $\frac{1}{2}$ with centre of enlargement (−1, −2).

Hint Work out the column vector from (−1, −2) to each vertex. Calculate the new vector, and then draw them from (−1, −2).

Exam-style question

③ Enlarge shape X by scale factor 1.5 with centre of enlargement (5, 6).

(3 marks)

Reflect Write a definition for the scale factor and the centre of enlargement.

3 Similarity and scale factors

When two shapes are similar, the ratio of the lengths of corresponding sides is equal to the scale factor of enlargement. When two shapes are similar, the corresponding angles are equal.

Guided practice

Rectangle ABCD is similar to rectangle PQRS.

Work out the length of rectangle PQRS.

A 7 cm B
2 cm [rectangle ABCD]
D   C

P   Q
3 cm [rectangle PQRS]
S   R

Worked exam question

Work out the scale factor of enlargement.

Scale factor = $\dfrac{3}{2}$ = ...

Multiply length AB by the scale factor.

PQ = 7 × .... = ... cm

The length of rectangle PQRS is .... cm.

You can also use the ratio of the sides of ABCD. The ratio of the width to the length is 2 : 7. Use an arrow diagram to work out the length of PQ.

$$\times 1\tfrac{1}{2} \left(\begin{array}{c} 2 : 7 \\ 3 : .... \end{array} \right) \times 1\tfrac{1}{2}$$

① Parallelograms PQRS and WXYZ are similar.

Complete these statements.

P   6 cm   Q
4 cm [parallelogram PQRS]
S   R

W 3 cm X
[parallelogram WXYZ]
Z   Y

a The side corresponding to PQ is ....

b Angle PSR is equal to angle ....

c The side corresponding to QR is ....

② **a** Work out the scale factor of the enlargement from parallelogram WXYZ to PQRS in Q1.

....

b Work out the scale factor of the enlargement from parallelogram PQRS to WXYZ.

....

c Work out the length WZ.

....

Hint Use an arrow diagram to compare the lengths of the corresponding sides.

③ Rectangle FGHJ is similar to rectangle KLMN.

Work out the width of rectangle KLMN. ....

F   12 cm   G
5 cm [rectangle FGHJ]
J   H

K   18 cm   L
[rectangle KLMN]
N   M

Exam-style question

④ Triangle PQR is similar to triangle XYZ.

Work out the two missing sides of triangle XYZ.

Q
12 cm / \ 18 cm
P / 27 cm \ R

Y
4 cm / \
X \ Z

....   (3 marks)

Reflect In your own words, explain what 'mathematically similar' means.

4 Describing enlargements with fractional scale factors

The centre of enlargement can be • on the shape • outside the shape • inside the shape.

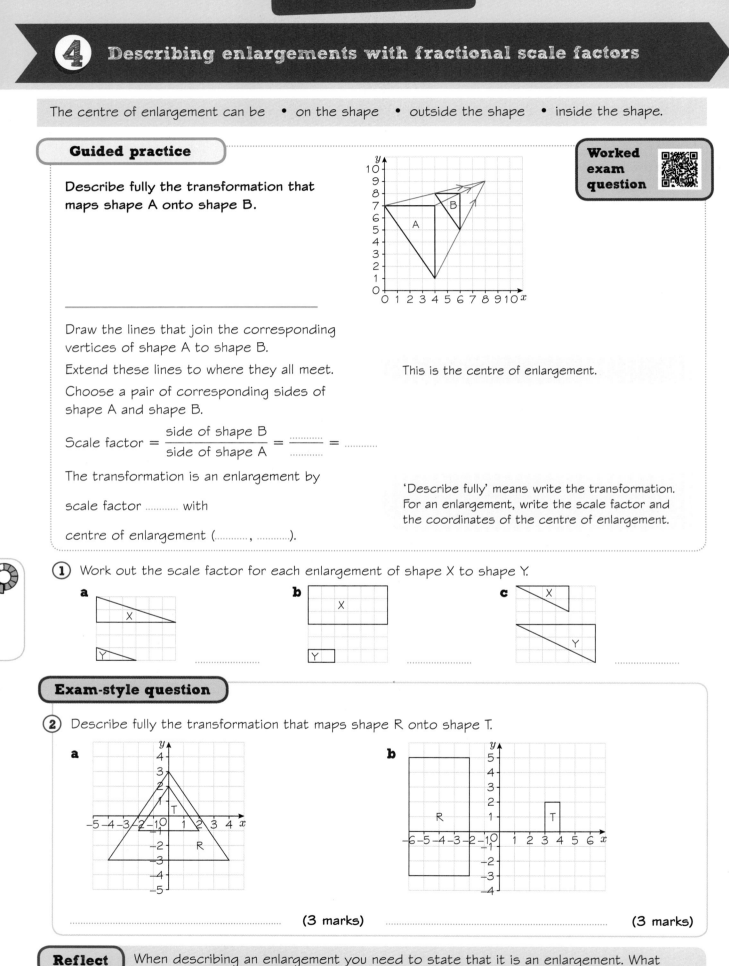

Guided practice

Describe fully the transformation that maps shape A onto shape B.

_____

Draw the lines that join the corresponding vertices of shape A to shape B.

Extend these lines to where they all meet.   This is the centre of enlargement.

Choose a pair of corresponding sides of shape A and shape B.

Scale factor $= \dfrac{\text{side of shape B}}{\text{side of shape A}} = \dfrac{\text{..........}}{\text{..........}} = \text{..........}$

The transformation is an enlargement by

scale factor .... with

centre of enlargement (.........., ...).

'Describe fully' means write the transformation. For an enlargement, write the scale factor and the coordinates of the centre of enlargement.

Worked exam question

① Work out the scale factor for each enlargement of shape X to shape Y.

a

b

c

....

....

....

Exam-style question

② Describe fully the transformation that maps shape R onto shape T.

a

b

.... (3 marks)

.... (3 marks)

Reflect When describing an enlargement you need to state that it is an enlargement. What other two pieces of information do you need to give?

Practise the methods

Answer this question to check where to start.

Check up

Match each translation from A to B with the correct vector.

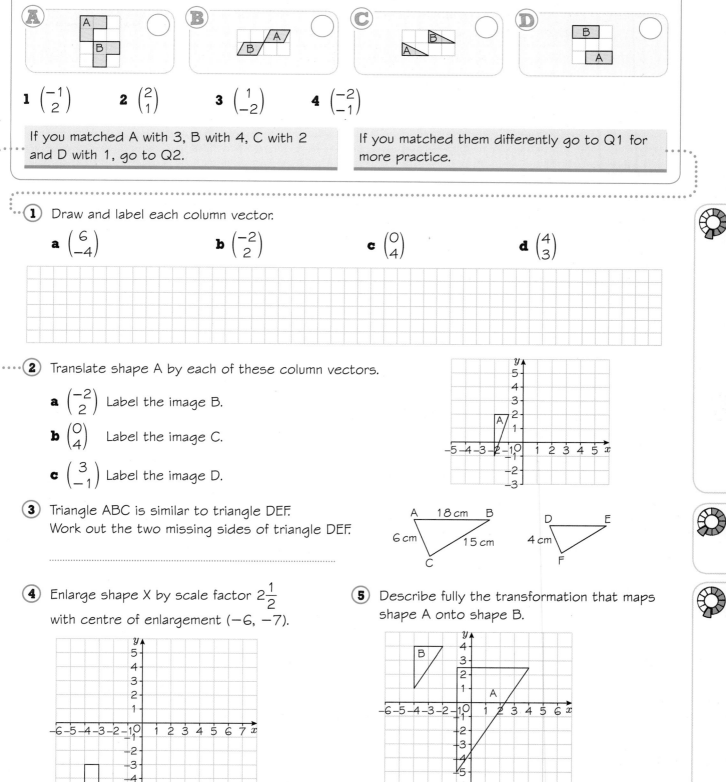

A ◯   **B** ◯   **C** ◯   **D** ◯

1 $\begin{pmatrix} -1 \\ 2 \end{pmatrix}$ **2** $\begin{pmatrix} 2 \\ 1 \end{pmatrix}$ **3** $\begin{pmatrix} 1 \\ -2 \end{pmatrix}$ **4** $\begin{pmatrix} -2 \\ -1 \end{pmatrix}$

If you matched A with 3, B with 4, C with 2 and D with 1, go to Q2.

If you matched them differently go to Q1 for more practice.

① Draw and label each column vector.

a $\begin{pmatrix} 6 \\ -4 \end{pmatrix}$ **b** $\begin{pmatrix} -2 \\ 2 \end{pmatrix}$ **c** $\begin{pmatrix} 0 \\ 4 \end{pmatrix}$ **d** $\begin{pmatrix} 4 \\ 3 \end{pmatrix}$

② Translate shape A by each of these column vectors.

a $\begin{pmatrix} -2 \\ 2 \end{pmatrix}$ Label the image B.

b $\begin{pmatrix} 0 \\ 4 \end{pmatrix}$ Label the image C.

c $\begin{pmatrix} 3 \\ -1 \end{pmatrix}$ Label the image D.

③ Triangle ABC is similar to triangle DEF.
Work out the two missing sides of triangle DEF.

④ Enlarge shape X by scale factor $2\frac{1}{2}$ with centre of enlargement $(-6, -7)$.

⑤ Describe fully the transformation that maps shape A onto shape B.

Problem-solve!

① A small photograph has a length of 5 cm and a
width of 4 cm. The photograph is enlarged to make
a large photograph with a width of 12 cm.

The two photographs are similar rectangles.
Work out the length of the large photograph. .... **(2 marks)**

② Triangle ABC is similar to triangle DEF.
Work out the perimeter of triangle DEF.

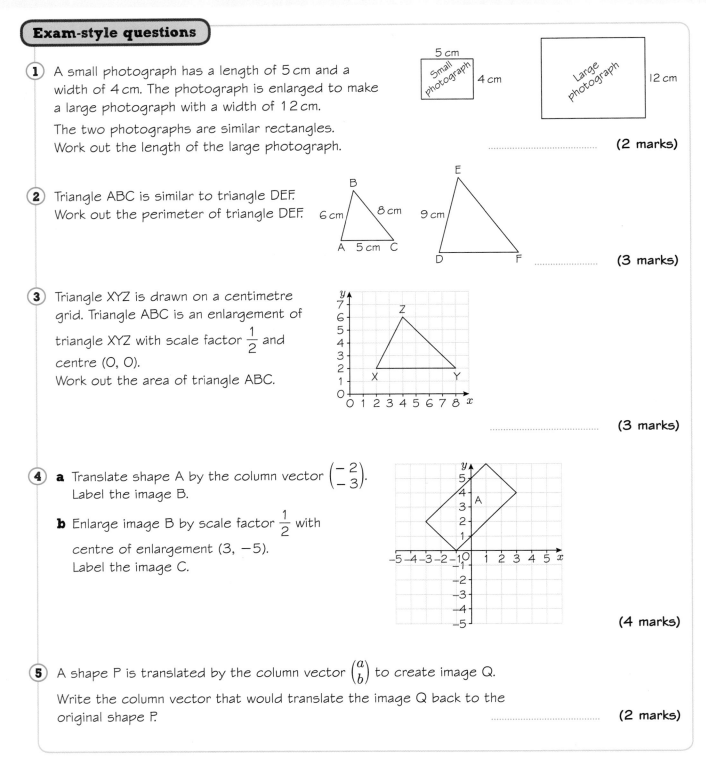

.... **(3 marks)**

③ Triangle XYZ is drawn on a centimetre
grid. Triangle ABC is an enlargement of
triangle XYZ with scale factor $\frac{1}{2}$ and
centre (0, 0).
Work out the area of triangle ABC.

.... **(3 marks)**

④ **a** Translate shape A by the column vector $\begin{pmatrix} -2 \\ -3 \end{pmatrix}$.
Label the image B.

b Enlarge image B by scale factor $\frac{1}{2}$ with
centre of enlargement (3, −5).
Label the image C.

(4 marks)

⑤ A shape P is translated by the column vector $\begin{pmatrix} a \\ b \end{pmatrix}$ to create image Q.

Write the column vector that would translate the image Q back to the
original shape P. .... **(2 marks)**

Now that you have completed this unit, how confident do you feel?

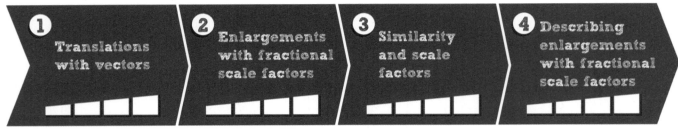

① Translations with vectors

② Enlargements with fractional scale factors

③ Similarity and scale factors

④ Describing enlargements with fractional scale factors

⑥ Averages

This unit will help you understand and use three different measures of average.

AO1 Fluency check

① For the data set 28, 33, 37, 30, 33, work out

    a the mean   **b** the median   **c** the mode.

    ....

② Round

    a 3.46 to 1 decimal place (d.p.)   **b** 42.7333 to 2 d.p.   **c** 13.999 to 3 significant figures (s.f.).

    ....

③ **Number sense**

    Work out

    a 25 + 46 + 27 + 33 + 34   **b** 350 ÷ 20   **c** 297 ÷ 3

    ....

Key points

To calculate the mean of a set of data, add up all the values and divide by the number of values.	To find the median of a set of data, put all the values in order and find the one in the middle.	The mode of a set of data is the most commonly occurring value. Not all sets of data have a mode.

These **skills boosts** will help you solve problems using different types of averages.

1 Estimating the mean from a grouped frequency table   **2** Which average is best?   **3** Calculating missing data values

You might have already done some work using averages. Before starting the first skills boost, rate your confidence on each process.

① The table shows the ages of members of a running club. Estimate the mean age.

Age, a (years)	Frequency
15 < a ≤ 25	7
25 < a ≤ 35	24
35 < a ≤ 45	41
45 < a ≤ 55	18
Total	

② The mean salary of staff in Mark's school is £22 000. What is a disadvantage of using the mean?

③ The mean of the five numbers 12, 15, 11, 13 and ☐ is 12. What is the missing number?

How confident are you?

1 Estimating the mean from a grouped frequency table

Guided practice

The table shows the heights of some bean plants one month after being planted. Estimate the mean height of the plants.
Give your answer to 1 d.p.

Height, h (cm)	Frequency
$0 < h \leq 10$	3
$10 < h \leq 20$	20
$20 < h \leq 30$	52
$30 < h \leq 40$	77
$40 < h \leq 50$	23

To find an estimate of the mean for grouped data use the midpoint of the class interval.

Height, h (cm)	Frequency	Midpoint of class	What the information means	Midpoint × frequency
$0 < h \leq 10$	3	5	Assume three plants each measure 5 cm	$3 \times 5 = 15$
$10 < h \leq 20$	20	15		
$20 < h \leq 30$	52			
$30 < h \leq 40$	77			
$40 < h \leq 50$	23			
Total				

Estimated total height of all bean plants = ...

Total number of bean plants = ...

Estimate of mean = ... ÷ .... = ...

Work out the total for the midpoint × frequency column.

Check the accuracy that the question asks for.

① The table shows the reaction times in a sports experiment.
Estimate the mean reaction time. Give your answer to 2 d.p.

Time, t (seconds)	Frequency	Midpoint of class	Midpoint × frequency
$0 < t \leq 0.5$	9		
$0.5 < t \leq 1.0$	13		
$1.0 < t \leq 1.5$	8		
$1.5 < t \leq 2.0$	2		
$2.0 < t \leq 2.5$	2		
Total			

Exam-style question

② Amir measures the heights of students in Year 11.
The table shows his results.
Estimate the mean height of the students.
Give your answer to 2 d.p.

Height, h (metres)	Frequency			
$1.55 < h \leq 1.6$	3			
$1.6 < h \leq 1.65$	14			
$1.65 < h \leq 1.7$	76			
$1.7 < h \leq 1.75$	81			
$1.75 < h \leq 1.8$	23			

.... **(4 marks)**

Reflect Explain why you can only estimate the mean for grouped data and not work out the exact mean.

Which average is best?

A game design company pays the following annual salaries.

CEO	Art director	Programmer	Designer	Assistant artist
£187 000	£62 000	£35 000	£45 000	£21 000

The mean salary is £70 000, the median salary is £45 000 and there is no mode.

Write any advantages and disadvantages for each of these averages.

Advantages/disadvantages of the mean are

....

....

> The mean is affected by extreme values.
> Are any of the salaries extreme values?

Advantages/disadvantages of the median are

....

....

> The median is not affected by extreme values, but it does not use all the values. Would it change if the CEO gave herself a pay rise?

Advantages/disadvantages of the mode are

....

....

> There may not always be a mode.

① Complete the table by writing an advantage and disadvantage of each average.

Hint Look at the statements in the Guided practice.

Average	Advantage	Disadvantage
Mean		
Median		
Mode		

② Choose the set of data for which the mean would be a good average to use. Explain your answer.

Hint You could work out the mean for each data set.

A 5.3, 4.6, 4.8, 5.5, 11.2, 4.5, 5.8, 5.5

B 78, 93, 87, 82, 90, 80, 85, 79, 88, 86

....

③ Choose the set of data for which the median would be a good average to use. Explain your answer.

A 5.3, 4.6, 4.8, 5.5, 11.2, 4.5, 5.8, 5.5

B 78, 93, 87, 82, 90, 80, 85, 79, 88, 86

....

④ Choose the set of data for which the mode would be the best average to use.
Explain your answer.

A 78, 93, 87, 82, 90, 80, 85, 79, 88, 86

B green, red, green, blue, blue, red, blue, yellow

....

⑤ Choose the most appropriate average to use for each set of data.
Explain your choice.

a Heights of Year 5 children (cm):

122, 135, 127, 129, 125, 131, 134, 124, 126

....

b Types of pets:

dog, rabbit, cat, dog, fish, dog, snake, rabbit

....

c Times measured in an experiment (s):

55.2, 64.3, 58.7, 56, 21.8, 60, 59.1

....

d Masses of 10p coins (g):

6.5, 6.3, 6.3, 6.4, 6.4, 6.3, 6.4, 6.4

....

Exam-style question

⑥ The waiting times, in minutes, for the first 10 patients in a dental surgery on Monday were

0   1   1   2   4   4   5   6   7   27

a Work out the mean waiting time for the patients.

.... **(2 marks)**

b Write a disadvantage of using the mean to find the average waiting time for the patients.

.... **(1 mark)**

c Which average would be the most appropriate to find the average waiting time for
the patients? Explain your answer.

.... **(2 marks)**

Reflect   Without looking at this page, write any advantages and disadvantages of each of the
three averages.

3 Calculating missing data values

You can use what you know about the mean, median and mode to find missing data values.

Guided practice

The mean of the numbers ?, 11, 7, 12 and 8 is 10.
Work out what the first number is.

Mean = total ÷ how many numbers

$\quad$ 10 = total ÷ 5 $\qquad\qquad\qquad\qquad$ Substitute the values you know.

Total = ... $\qquad\qquad\qquad$ The mean is 10 and there are 5 numbers.

$\quad$? = 50 − .... = ... $\qquad$? + 11 + 7 + 12 + 8 = total = ...

The first number is ....

(1) Complete and use the reverse function machine to work out the total for a set of six numbers with a mean of 15.

total → ÷ how many numbers → mean

total ← ⬡ ← mean

....

(2) The mean of 19, 24, 23, 18, 20, 23, 19 and x is 21.
Work out the value of x. $\qquad\qquad$ ....

Exam-style question

(3) The mean of 145, 126, 130 and c is 137.
Work out the value of c. $\qquad\qquad$ .... **(2 marks)**

(4) Write five different numbers with a mean of 11. $\qquad$ ....

(5) Write six numbers with a mean of 15.
Two of the numbers must be the same and
the other four must be different. $\qquad\qquad$ ....

(6) The median of the five numbers 9, 11, 7, 10 and x is 9 and the range is 4.
What values could x be? $\qquad\qquad$ ....

Hint Put the numbers in order. Start by working out the median and range for the four numbers given.

(7) Write seven numbers with a mode of 12, a median of 12
and a range of 8. $\qquad\qquad$ ....

Reflect

Without looking at this page, write the rule for working out the total of the data values when you know the mean and how many values there are.

Practise the methods

Answer this question to check where to start.

Check up

Tick the correct midpoint of the class interval $35 < t \leqslant 40$.

A 37 ◯   **B** 38 ◯   **C** 37.5 ◯   **D** 35 ◯

If you ticked C that is correct. Now write possible class intervals for the other midpoints then go to Q2.

If you ticked A, B or D go to Q1 for more practice.

① Work out the midpoint for each class interval.

   a $20 < l \leqslant 25$ ....   **b** $75 < m \leqslant 80$ ....   **c** $15 < p \leqslant 18$ ....

Exam-style question

② The table shows the weights of babies born in a hospital during June. Estimate the mean weight of the babies to 2 d.p.

Weight, w (kg)	Frequency			
$2 < w \leqslant 2.5$	4			
$2.5 < w \leqslant 3$	38			
$3 < w \leqslant 3.5$	73			
$3.5 < w \leqslant 4$	47			
$4 < w \leqslant 4.5$	12			

   .... **(4 marks)**

③ Choose the most appropriate average to use for each set of data. Explain your choice.

   a elm, beech, oak, elm, oak, elm, hawthorn

   ....

   b 46, 50, 49, 50, 45, 44, 45, 125, 47

   ....

   c 22.5, 22.7, 21.3, 23.6, 22.3, 21.9, 23.2

   ....

④ Write six different numbers with a mean of 14. ....

⑤ Six cards show the numbers 17, 10, 15, 21, 11 and p.

   a Work out the value of p if the mean is 16. ....

   b Work out a possible value of p if the median is 14. ....

Problem-solve!

1 Here are five number cards.

$$8 \quad \boxed{} \quad 5 \quad 9 \quad 5$$

One of the cards is turned over so you cannot see the number on it.
The mean of the five numbers is 7.
Work out the number you cannot see. .... **(3 marks)**

2 Each of these cards has a number on it. The numbers are hidden.

$$? \quad ? \quad ?$$

The mode of the three numbers is 7. The mean of the three numbers is 8.
Work out the three numbers on the cards. .... **(2 marks)**

3 28 students in Class A did a science exam. 32 students in Class B did the same exam.
The mean mark for the 28 students in Class A was 77.5.
The mean mark for all 60 students was 81.2.
Work out the mean mark for the students in Class B.
Give your answer to 1 d.p. .... **(3 marks)**

4 There are 12 boys and 18 girls in Mrs Ball's class. Mrs Ball gave all the class a test.
The mean mark for all the class was 65.
The mean mark for the girls was 63.
Work out the mean mark for the boys. .... **(3 marks)**

5 The chart shows the
distribution of times taken by
Year 11 students to travel
to school.
Work out an estimate of the
mean time taken.
Round your answer to
1 d.p. .... **(4 marks)**

Now that you have completed this unit, how confident do you feel?

1 Estimating the mean from a grouped frequency table

2 Which average is best?

3 Calculating missing data values

⑦ Probability

This unit will help you draw and use Venn and tree diagrams to work out probabilities.

AO1 Fluency check

① List

    a the first eight square numbers   **b** the first eight prime numbers **c** the factors of 12.

    1 4 9 16 25 36 42 64   2 3 5 11 13 17 19 23 29 31 1, 2, 3, 4, 6, 12

② **Number sense**

Work out

    a $\dfrac{3}{4} \times \dfrac{1}{2} =$ $\dfrac{3}{8}$ **b** $\dfrac{4}{7} \times \dfrac{5}{8} =$ $\dfrac{20}{56}$ **c** $\dfrac{3}{10} + \dfrac{9}{10} =$ $\dfrac{12}{10}$ **d** $\dfrac{2}{3} + \dfrac{2}{9} =$ $\dfrac{24}{27}$

Key points

If one event does not affect the outcome of another, the two events are independent, for example rolling two dice.

If one event depends on the outcome of another, the two events are dependent, for example picking two chocolates from a box without replacement.

These **skills boosts** will help you use diagrams to work out probabilities.

1 Venn diagrams   >   **2** Tree diagrams for independent events   >   **3** Tree diagrams for dependent events   >   **4** More tree diagrams

You might have already done some work on Venn and tree diagrams. Before starting the first skills boost, rate your confidence using these diagrams.

① Complete the Venn diagram using the numbers 1 to 10. Use it to work out the probability of a number being a multiple of 2 and a factor of 10.

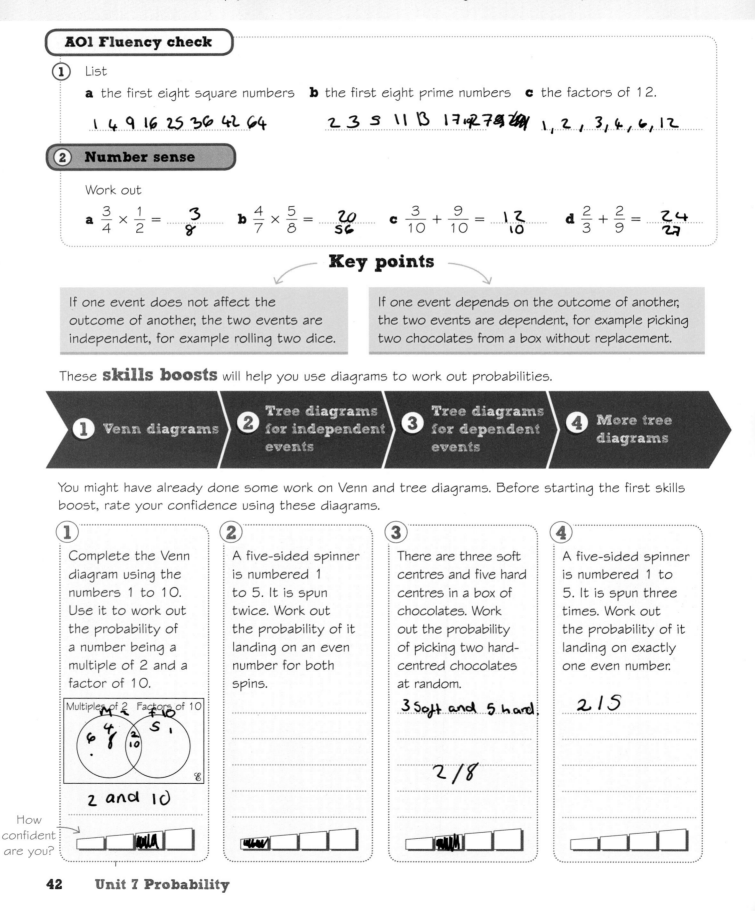

2 and 10

② A five-sided spinner is numbered 1 to 5. It is spun twice. Work out the probability of it landing on an even number for both spins.

③ There are three soft centres and five hard centres in a box of chocolates. Work out the probability of picking two hard-centred chocolates at random.

3 Soft and 5 hard.

2/8

④ A five-sided spinner is numbered 1 to 5. It is spun three times. Work out the probability of it landing on exactly one even number.

2 / 5

How confident are you?

1 Venn diagrams

A Venn diagram is a way of showing sets.

Guided practice

Marton High School teaches two languages: French and Spanish.
There are 253 students in Year 11.
41 students study both French and Spanish.
102 students study French.
12 students do not study a language at all.

a Draw a Venn diagram to show this data.

A Year 11 student is picked at random.
Work out the probability that the student studies

b both French and Spanish

c neither French nor Spanish

d Spanish.

---

a Use the information given in the question to complete the Venn diagram.

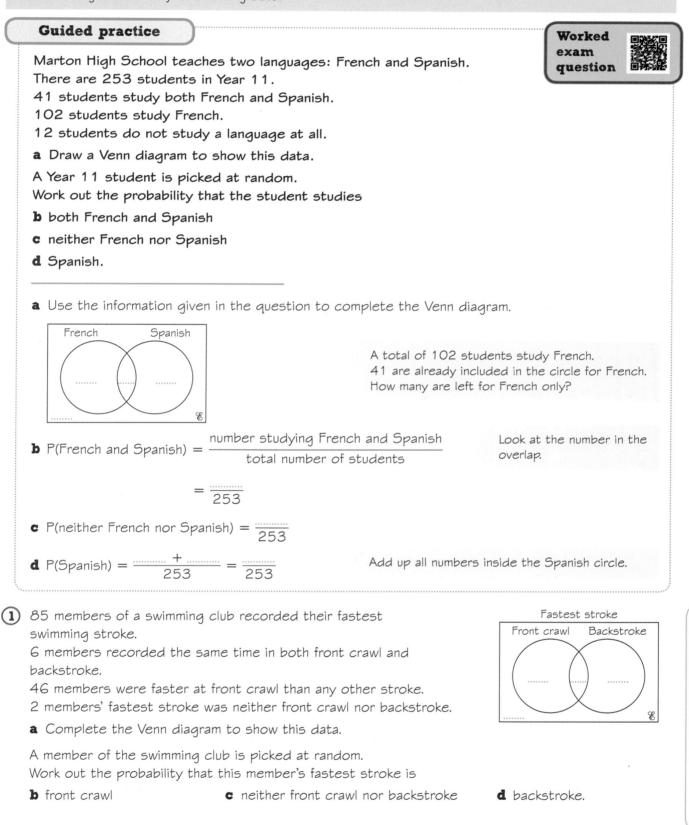

A total of 102 students study French.
41 are already included in the circle for French.
How many are left for French only?

b P(French and Spanish) = $\dfrac{\text{number studying French and Spanish}}{\text{total number of students}}$

Look at the number in the overlap.

$= \dfrac{.........}{253}$

c P(neither French nor Spanish) = $\dfrac{.........}{253}$

d P(Spanish) = $\dfrac{......... + ...}{253} = \dfrac{.........}{253}$

Add up all numbers inside the Spanish circle.

---

① 85 members of a swimming club recorded their fastest swimming stroke.
6 members recorded the same time in both front crawl and backstroke.
46 members were faster at front crawl than any other stroke.
2 members' fastest stroke was neither front crawl nor backstroke.

a Complete the Venn diagram to show this data.

A member of the swimming club is picked at random.
Work out the probability that this member's fastest stroke is

b front crawl

c neither front crawl nor backstroke

d backstroke.

Exam-style question

(2) A sandwich bar records the sandwiches it sells on Saturday.

55 sandwiches are sold in total.

13 have a ham and cheese filling.

35 have a ham filling.

14 have neither a ham nor cheese filling.

a Complete the Venn diagram to show this data.

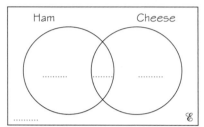

(3 marks)

A customer is picked at random.

Work out the probability that this customer orders

b a ham and cheese sandwich .... (1 mark)

c a cheese sandwich. .... (1 mark)

(3) The manager of a health club wants to know if members use the gym or the swimming pool.

The health club has 225 members in total.

73 members use both the gym and the swimming pool.

156 members use the gym.

123 members use the swimming pool.

a Draw a Venn diagram to show this data.

A member is picked at random. Work out the probability that they use

b only the gym ....

c only the swimming pool ....

d neither the gym nor the swimming pool. ....

Reflect   Work out the total of the numbers in each of your Venn diagrams. Does this total match any of the numbers given in each question?

2 Tree diagrams for independent events

A tree diagram can help you work out the combined probabilities of more than one event.

Guided practice

Worked exam question

A five-sided spinner is numbered 1 to 5. It is spun twice.

a Complete the probability tree diagram.

Work out the probability of

b both spins being odd numbers

c spinning one even and one odd number.

1st spin   2nd spin

Odd
— Odd
— Even

Even
— Odd
— Even

a

1st spin   2nd spin

$\frac{3}{5}$ Odd
$\frac{3}{5}$ — Odd
.... — Even

$\frac{2}{5}$ Even
.... — Odd
.... — Even

There are three odd numbers and two even numbers on the spinner.

Work out $P(\text{odd}) = \dfrac{\text{..........}}{5}$

$P(\text{even}) = \dfrac{\text{..........}}{5}$

b $P(\text{both odd}) = \dfrac{3}{5} \times \dfrac{3}{5} = \dfrac{\text{..........}}{\text{..........}}$

Go along the branches 'Odd', 'Odd' and multiply the two probabilities.

c $P(\text{one even and one odd}) = \dfrac{3}{5} \times \dfrac{2}{5} + \dfrac{\text{........}}{\text{..........}} \times \dfrac{\text{..........}}{\text{..........}}$

$= \dfrac{\text{..........}}{\text{..........}} + \dfrac{\text{..........}}{\text{..........}}$

$= \dfrac{\text{..........}}{\text{..........}}$

To work out the probability of spinning an even number and an odd number, add together the probabilities for 'Odd', 'Even' and 'Even', 'Odd'.

① A six-sided spinner is numbered 1 to 6. It is spun twice.

a Complete the probability tree diagram.

Work out the probability that

b both results are not multiples of 3

....

c only one result is a multiple of 3.

....

1st spin   2nd spin

$\frac{2}{6} = \frac{1}{3}$ Multiple of 3
$\frac{1}{3}$ Multiple of 3
.... Not a multiple of 3

.... Not a multiple of 3
.... Multiple of 3
.... Not a multiple of 3

② There are 3 red and 5 blue counters in a bag.
Emily chooses a counter at random.
She makes a note of the colour and then puts it back in the bag.
She then picks a second counter at random from the bag.

   a Complete the probability tree diagram.

   Work out the probability that

   b both counters are blue ....

   c one counter is red and the other is blue. ....

1st counter   2nd counter

$\frac{3}{8}$ ⟨ Red .... Red / .... Blue

.... Blue ⟨ .... Red / .... Blue

③ At a school fundraising event there is a hook-a-duck game.
There are 10 ducks.
One of the ducks is marked underneath with a cross.
If someone hooks the duck with the cross they win a prize.
Karl pays for two games.
He catches a duck, looks to see if it has a cross and then puts it
back on the water before he hooks another duck.

   a Complete the probability tree diagram.

   Work out the probability that Karl

   b doesn't win any prizes ....

   c wins exactly one prize. ....

1st duck   2nd duck

.... X ⟨ .... X / .... No X

.... No X ⟨ .... X / .... No X

④ Felix uses the letters of his name to make a spinner.
The probability of the spinner landing on a vowel is 0.4.
The probability of it landing on a consonant is 0.6.
Felix spins the spinner to see if it lands on a vowel or a consonant.
He spins the spinner twice.

   a Complete the probability tree diagram.

   Work out the probability that the spinner lands on

   b two consonants ....

   c one consonant and one vowel. ....

1st letter   2nd letter

.... V .... ⟨ .... V .... / .... C ....

.... C .... ⟨ .... V .... / .... C ....

Exam-style question

⑤ Ten cards numbered 1 to 10 are shuffled and placed face down.
A card is picked at random to see if it is a multiple of 4 or not before being replaced.
The cards are then shuffled and placed face down before another card is picked.

   a Complete the probability tree diagram.

   Work out the probability that

   b the first card is a multiple of 4 and
   the second card isn't

   c exactly one card is a multiple of 4.

1st card   2nd card

.... multiple of 4 ⟨ .... multiple of 4 / .... not a multiple of 4

.... not a multiple of 4 ⟨ .... multiple of 4 / .... not a multiple of 4

(1 mark)

.... (1 mark)

.... (2 marks)

Reflect Write a sentence to explain what independent events are.

46   Unit 7 Probability

3 Tree diagrams for dependent events

If one event depends upon the outcome of another event, the two events are dependent events. The probability of the second event depends on the outcome of the first event.

Guided practice

Jane has a box of chocolates.
4 of the chocolates have soft centres
and 3 have hard centres.
Jane picks a chocolate at random and eats it.
Then she picks another.

a Complete the probability tree diagram.

Work out the probability that Jane picks

b two hard centres

c one of each kind of chocolate.

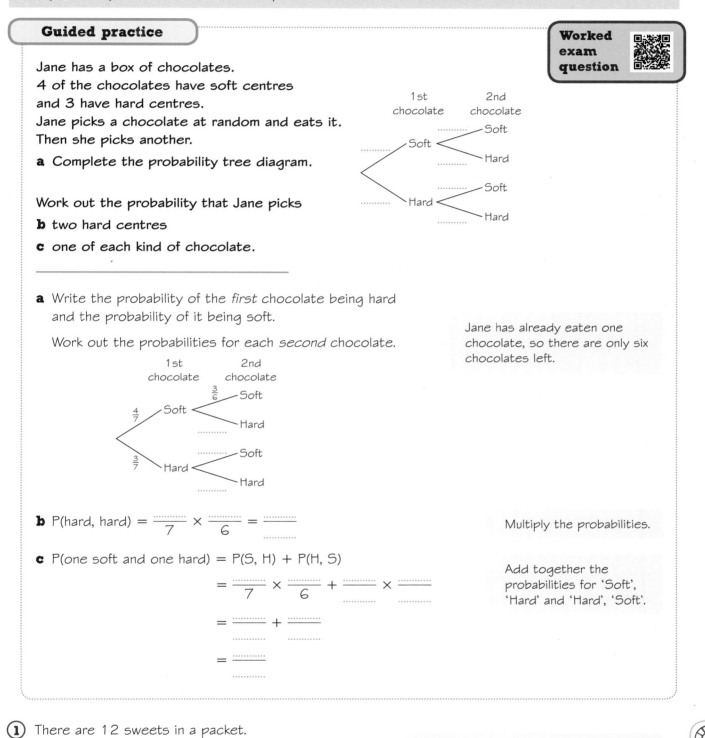

1st chocolate   2nd chocolate

Soft ⟨ Soft / Hard

Hard ⟨ Soft / Hard

a Write the probability of the *first* chocolate being hard and the probability of it being soft.

Work out the probabilities for each *second* chocolate.

Jane has already eaten one chocolate, so there are only six chocolates left.

1st chocolate   2nd chocolate

$\frac{4}{7}$ Soft ⟨ $\frac{3}{6}$ Soft / .... Hard

$\frac{3}{7}$ Hard ⟨ .... Soft / .... Hard

b $P(\text{hard, hard}) = \dfrac{..........}{7} \times \dfrac{..........}{6} = \dfrac{..........}{..........}$

Multiply the probabilities.

c $P(\text{one soft and one hard}) = P(S, H) + P(H, S)$

$= \dfrac{..........}{7} \times \dfrac{..........}{6} + \dfrac{..........}{..........} \times \dfrac{..........}{..........}$

Add together the probabilities for 'Soft', 'Hard' and 'Hard', 'Soft'.

$= \dfrac{..........}{..........} + \dfrac{..........}{..........}$

$= \dfrac{..........}{..........}$

① There are 12 sweets in a packet.

7 of the sweets are strawberry flavoured and the other 5 are orange flavoured.

Toby picks a strawberry sweet and eats it.
He then picks another sweet at random.

What is the probability that this sweet is also strawberry? ....

Hint There were 7 strawberry flavoured sweets to start with. After eating one, how many are left?

(2) There are 8 cartons in the fridge.

5 are orange juice and 3 are apple juice.

Amelia picks two cartons at random.

a Complete the probability tree diagram.

Hint Picking 2 cartons at random is the same as picking one carton at random and then a second carton at random.

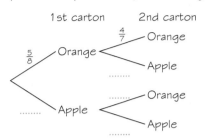

1st carton   2nd carton

$\frac{5}{8}$ Orange

$\frac{4}{7}$ Orange

.... Apple

Apple

.... Orange

.... Apple

b Work out the probability that both cartons are the same.

Hint P(both cartons are the same) = P(O, O) + P(A, A)

....

Exam-style question

(3) Adil has seven blue and three black pens in his pencil case. He picks two pens at random.

a Complete the probability tree diagram.

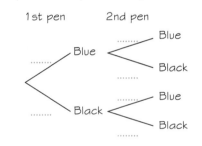

1st pen   2nd pen

.... Blue

Blue

.... Black

....

.... Blue

Black

.... Black

(2 marks)

Work out the probability that Adil picks

b two black pens   .... **(1 mark)**

c one pen of each colour.   .... **(2 marks)**

---

 (4) There are 11 red counters and 7 blue counters in a bag.

Dalia picks two counters at random.

a Complete the probability diagram.

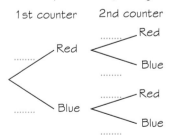

1st counter   2nd counter

.... Red

Red

.... Blue

....

.... Red

Blue

.... Blue

b Work at the probability that both counters are the same colour.

....

Reflect   Write a sentence to explain what dependent events are.

4 More tree diagrams

Jack has a box of muffins. 4 of the muffins are blueberry,
3 are double chocolate chip and 2 are white chocolate.
Jack picks 2 muffins at random.

a Complete the probability tree diagram.

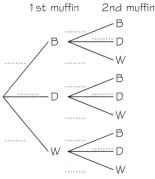

b Work out the probability that the 2 muffins
Jack picks are the same.

_____

a Work out the probability for each branch. The two events are dependent.

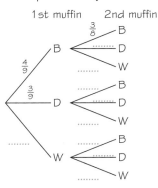

There are three types of muffins so three
branches are needed. Only two muffins
are being picked so only two columns of
branches are needed.

If the first muffin is blueberry, the probability
that the second muffin can be blueberry is
$\frac{3}{8}$ because there are 3 blueberry muffins left
and 8 muffins left in total.

b P(same) = P(B, B) + P(D, D) + P(W, W)

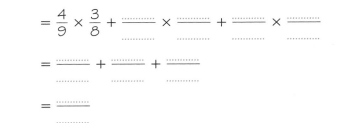

$$= \frac{4}{9} \times \frac{3}{8} + \frac{\dots}{\dots} \times \frac{\dots}{\dots} + \frac{\dots}{\dots} \times \frac{\dots}{\dots}$$

$$= \frac{\dots}{\dots} + \frac{\dots}{\dots} + \frac{\dots}{\dots}$$

$$= \frac{\dots}{\dots}$$

① Erin picks two chocolates at random from a box containing 5 milk, 3 dark and
2 white chocolates. Circle the correct shaped tree diagram for this situation.

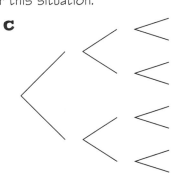

A   **B**   **C**

Hint Erin picks two chocolates so there
should be two columns of branches.

(2) **a** Complete Erin's probability tree diagram from Q1.

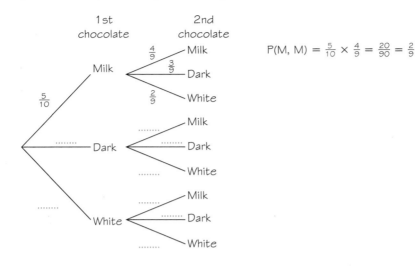

$P(M, M) = \frac{5}{10} \times \frac{4}{9} = \frac{20}{90} = \frac{2}{9}$

b Work out the probability that Erin picks two different chocolates. ....

Hint P(two different chocolates) = P(M, D) + P(M, W) + P(D, M) + P(D, W) + P(W, M) + P(W, D)

Exam-style question

(3) Two fair six-sided dice are rolled.

a Complete the probability tree diagram.

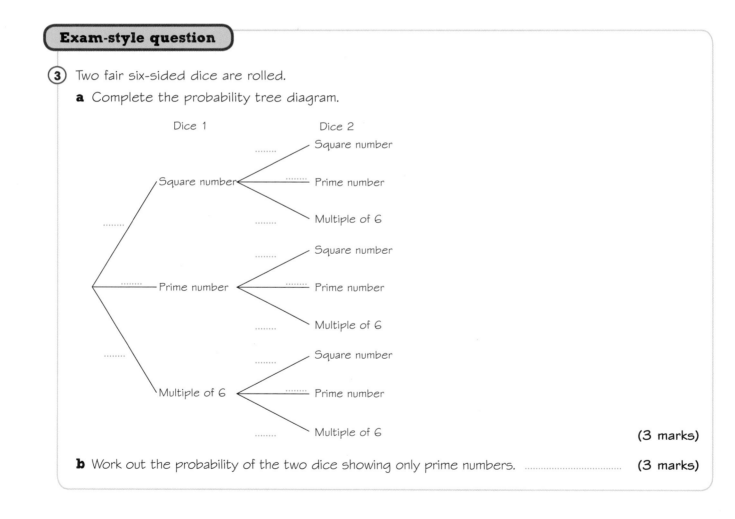

(3 marks)

b Work out the probability of the two dice showing only prime numbers. .... (3 marks)

Reflect What information in each question on this page tells you if the events are independent or dependent?

Practise the methods

Answer this question to check where to start.

Tick the correct calculation to work out the probability of both numbers being odd.

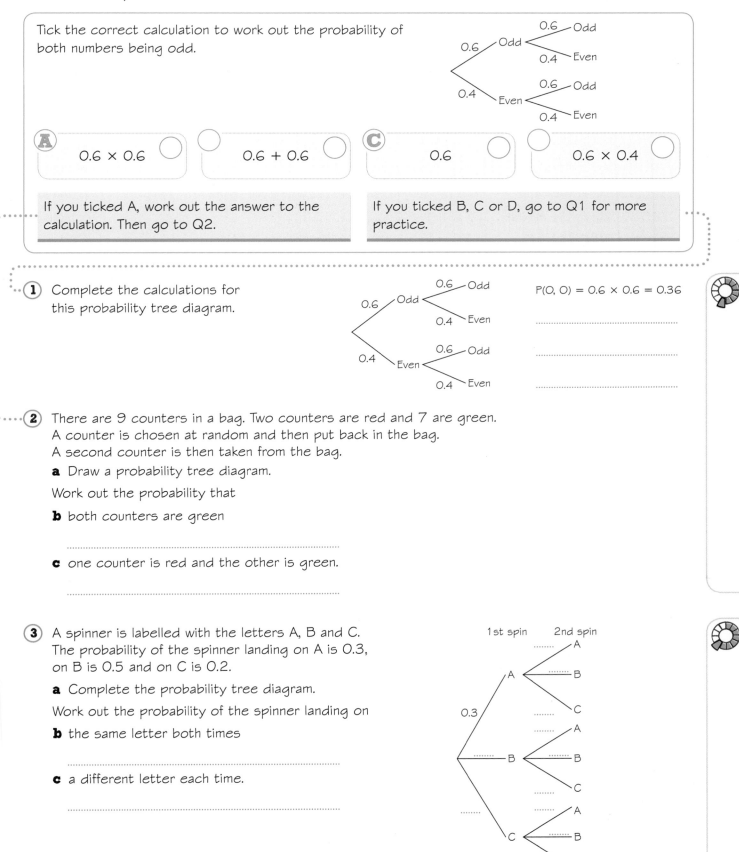

0.6 Odd
0.6 Odd
0.4 Even
0.4 Even
0.6 Odd
0.6 Odd
0.4 Even

A 0.6 × 0.6 ◯   ◯ 0.6 + 0.6 ◯   **C** 0.6 ◯   ◯ 0.6 × 0.4 ◯

If you ticked A, work out the answer to the calculation. Then go to Q2.

If you ticked B, C or D, go to Q1 for more practice.

1 Complete the calculations for this probability tree diagram.

0.6 Odd
0.6 Odd
0.4 Even
0.4 Even
0.6 Odd
0.6 Odd
0.4 Even

P(O, O) = 0.6 × 0.6 = 0.36

....

....

....

2 There are 9 counters in a bag. Two counters are red and 7 are green.
A counter is chosen at random and then put back in the bag.
A second counter is then taken from the bag.

   a Draw a probability tree diagram.

   Work out the probability that

   b both counters are green

   ....

   c one counter is red and the other is green.

   ....

3 A spinner is labelled with the letters A, B and C.
The probability of the spinner landing on A is 0.3,
on B is 0.5 and on C is 0.2.

   a Complete the probability tree diagram.

   Work out the probability of the spinner landing on

   b the same letter both times

   ....

   c a different letter each time.

   ....

1st spin   2nd spin
.... A
A .... B
0.3 C
.... A
B .... B
.... C
.... A
C .... B
.... C

Problem-solve!

1 **a** Complete the Venn diagram using the numbers 1 to 20.

Square numbers Even numbers

ℰ

(3 marks)

A number is selected at random. Work out the probability that the number is

b a square number .... (1 mark)

c an even number .... (1 mark)

d an even square number .... (1 mark)

e neither an even number nor a square number. .... (2 marks)

2 Isaac plays two tennis matches.
The probability of him winning a match is 0.7.

a Complete the probability tree diagram.

1st match   2nd match

0.7 — Win .... Win
.... Lose
.... Lose — Win
.... Lose

(2 marks)

Work out the probability of Isaac

b winning both matches .... (1 mark)

c losing *at least* one match. .... (2 marks)

3 Jordan has ten socks.
Six of the socks are black.
Four of the socks are green.
Jordan takes two of the socks at random.
Work out the probability that both socks are the same colour. .... (4 marks)

4 There are ten cakes on a plate: one doughnut, six cupcakes and three brownies.
Karis takes a cake at random and eats it.
Then she takes a second cake at random.
Work out the probability that Karis takes two different types of cake. .... (4 marks)

Now that you have completed this unit, how confident do you feel?

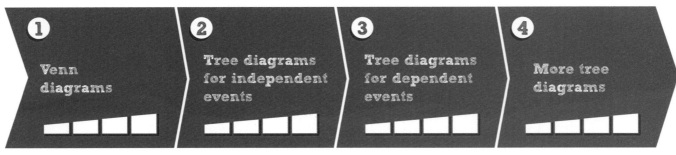

1 Venn diagrams

2 Tree diagrams for independent events

3 Tree diagrams for dependent events

4 More tree diagrams

⑧ Scatter graphs

This unit will help you describe correlation, use a line of best fit and understand interpolation and extrapolation.

AO1 Fluency check

① Draw crosses on each set of axes to show the type of correlation.

a positive correlation   b negative correlation   c no correlation

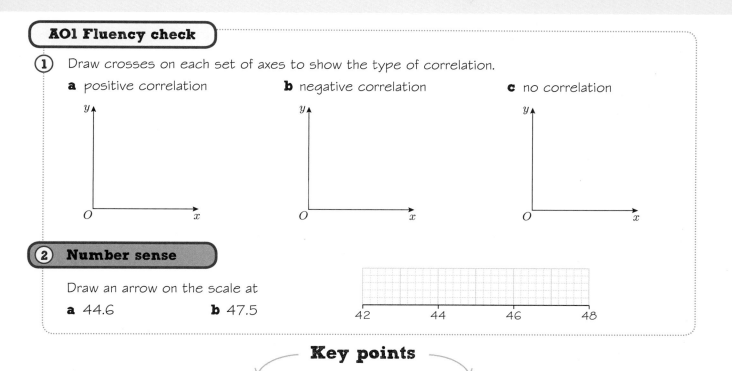

② **Number sense**

Draw an arrow on the scale at

a 44.6   b 47.5

Key points

A scatter graph displays the relationship between two sets of data.

Each data set is represented by a variable.

These **skills boosts** will help you interpret correlation and use lines of best fit.

> ❶ **Interpreting scatter graphs and correlation**
> ❷ **Using the line of best fit**

You might have already done some work on scatter graphs. Before starting the first skills boost, rate your confidence on each topic.

①
Mr Jones measures the height and foot length of his students. Describe the relationship you would expect him to find between height and foot length.

②
Dan measures the height and circumference of some trees. He shows his data in a scatter graph and draws a line of best fit. The tallest tree is 7.3 metres tall.

He finds another tree 13 metres tall. He decides to use his line of best fit to estimate the tree's circumference. Explain why this will not provide a reliable estimate.

How confident are you?

1 Interpreting scatter graphs and correlation

The relationship between two sets of data is called correlation.

Guided practice

The scatter graph shows the length and wingspan of various types of British birds.

a Describe the relationship between the length and wingspan of the birds.

b State the type of correlation between the length and wingspan of the birds.

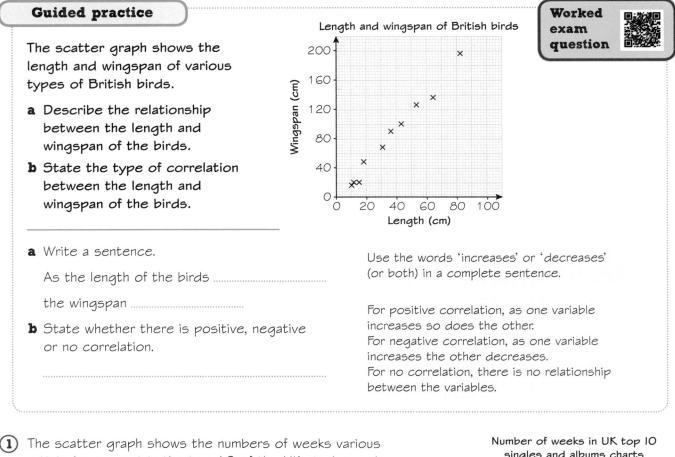

Length and wingspan of British birds

Worked exam question

a Write a sentence.

As the length of the birds ....

the wingspan ....

b State whether there is positive, negative or no correlation.

....

Use the words 'increases' or 'decreases' (or both) in a complete sentence.

For positive correlation, as one variable increases so does the other.
For negative correlation, as one variable increases the other decreases.
For no correlation, there is no relationship between the variables.

---

1 The scatter graph shows the numbers of weeks various artists have spent in the top 10 of the UK singles and albums charts.

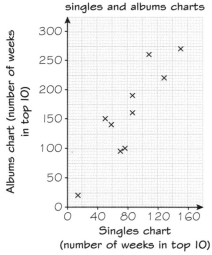

Number of weeks in UK top 10 singles and albums charts

a Describe the relationship between the numbers of weeks spent in the top 10 of the UK singles and albums charts.

....

....

b State the type of correlation between the numbers of weeks spent in the top 10 of the UK singles and albums charts.

....

2 The scatter graph shows the engine sizes of cars from one manufacturer and how many seconds it takes the cars to go from 0 to 60 mph.

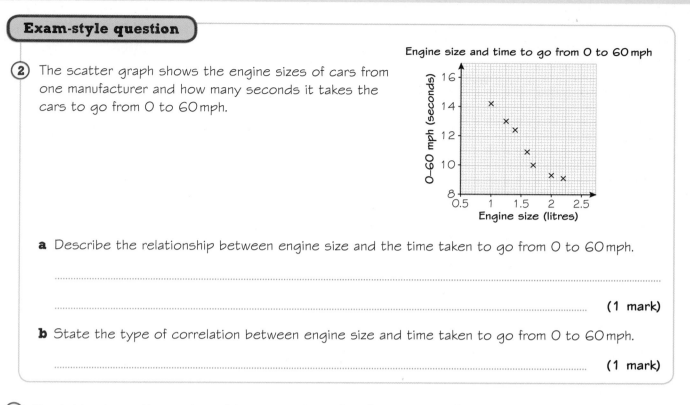

Engine size and time to go from 0 to 60 mph

a Describe the relationship between engine size and the time taken to go from 0 to 60 mph.

....

.... **(1 mark)**

b State the type of correlation between engine size and time taken to go from 0 to 60 mph.

.... **(1 mark)**

3 The table shows the number of hours spent on the phone per week by 15 people, and their salaries.

Phone hours per week	8	2	7	9	10	1	3	6	9	5	4	6	1	8	3
Salary (£1000s)	52	25	46	60	66	17	21	34	57	33	25	48	15	27	31

a Plot the data on the grid.

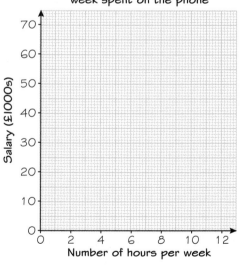

Salary and number of hours per week spent on the phone

b Describe the relationship and the correlation between the number of hours spent on the phone per week and salary.

....

....

c Is the relationship between the two variables causal?

....

Hint Do you think one variable directly affects the other?

Reflect Does correlation between two variables mean that a change in one variable causes a change in the other? Explain your answer.

2 Using the line of best fit

Interpolation is using the line of best fit to predict data values within the range of the given data.
Extrapolation is using the line of best fit to predict data values outside the range of the given data,
so is less reliable and may give impossible estimates.

Guided practice

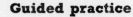
The scatter graph shows the exam results
for a group of students.
The maximum number of marks on each
paper is 60.

a Paul got 43 marks on Paper 1
but was absent for Paper 2.
Estimate the mark Paul would have got
for Paper 2.

b Katie got the maximum of 60 marks
on Paper 2 but was absent for Paper 1.
Estimate the mark Katie would have got on Paper 1.

c Which is the more reliable estimate? Explain your answer.

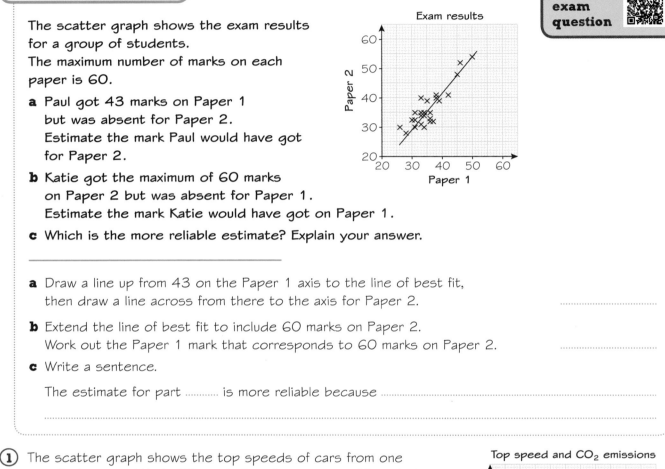

a Draw a line up from 43 on the Paper 1 axis to the line of best fit,
then draw a line across from there to the axis for Paper 2.

b Extend the line of best fit to include 60 marks on Paper 2.
Work out the Paper 1 mark that corresponds to 60 marks on Paper 2.

c Write a sentence.

The estimate for part .... is more reliable because ....

1 The scatter graph shows the top speeds of cars from one
manufacturer, and their CO_2 emissions in grams per kilometre.

a The manufacturer designs a car with a top speed of
140 mph. Estimate the CO_2 emissions of this car.

b The manufacturer designs another car with a top speed
of 110 mph. Estimate the CO_2 emissions of this car.

c Which is the more reliable estimate, your answer to
part a or part b? Explain why.

Reflect Explain the difference between interpolation and extrapolation.

Practise the methods

Answer this question to check where to start.

Check up

Match each pair of variables to the correct descriptions.

1 Height and arm span

2 Shoe size and hair length

3 Exercise and resting heart rate

4 Height and hours of TV watched

(A) One variable increases as the other increases.

(B) One variable increases as the other decreases.

(C) There is no relationship between the variables.

(D) One change causes the other.

If you matched 1 with A, 2 with C, 3 with B and D, and 4 with C, describe the correlation for each set of data. Then go to Q2.

If you matched them differently, go to Q1 for more practice.

1 Match each real-life example to the correct type of relationship and state the correlation.

a Distance walked and tread left on shoes

b Eye colour and age

c Arm length and hand span

A One variable increases as the other increases.

B One variable increases as the other decreases.

C There is no relationship between the variables.

2 The scatter graph shows the cost and top speed of cars from one manufacturer.

a Describe the relationship between the cost and top speed of the cars. ....

b State the type of correlation between the cost and top speed of the cars. ....

c You should not use a line of best fit to predict the cost of a car with a top speed of 140 mph. Give one reason why not.

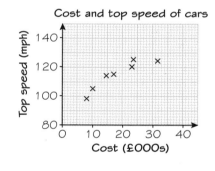

Cost and top speed of cars

Exam-style question

3 The scatter graph shows the marks Mrs Brown's class got in their Italian listening and reading exams.

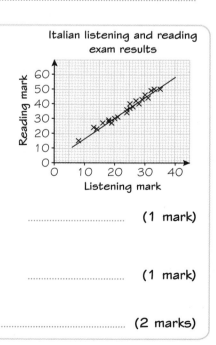

Italian listening and reading exam results

a Seema's mark in the reading exam was 41 but she was absent for the listening exam.
Estimate the mark she would have got in the listening exam. .... **(1 mark)**

b Karl got 40 marks in the listening exam but was absent for the reading exam.
Estimate the mark Karl would have got in the reading exam. .... **(1 mark)**

c Which is the more reliable estimate? Explain your answer.

.... **(2 marks)**

Problem-solve!

① The scatter graph shows information about 14 European countries. For each country, it shows the wealth (GDP per capita) and the life expectancy.

a Describe the relationship between the wealth and the life expectancy of these countries.

....

.... (1 mark)

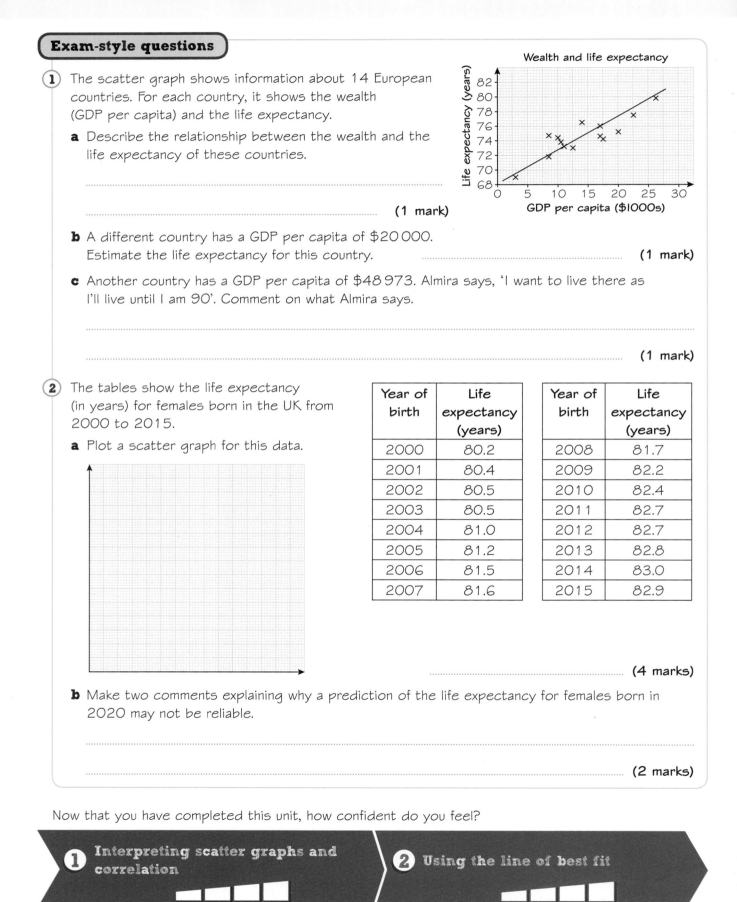

Wealth and life expectancy

b A different country has a GDP per capita of $20 000. Estimate the life expectancy for this country. .... (1 mark)

c Another country has a GDP per capita of $48 973. Almira says, 'I want to live there as I'll live until I am 90'. Comment on what Almira says.

....

.... (1 mark)

② The tables show the life expectancy (in years) for females born in the UK from 2000 to 2015.

a Plot a scatter graph for this data.

Year of birth	Life expectancy (years)	Year of birth	Life expectancy (years)
2000	80.2	2008	81.7
2001	80.4	2009	82.2
2002	80.5	2010	82.4
2003	80.5	2011	82.7
2004	81.0	2012	82.7
2005	81.2	2013	82.8
2006	81.5	2014	83.0
2007	81.6	2015	82.9

.... (4 marks)

b Make two comments explaining why a prediction of the life expectancy for females born in 2020 may not be reliable.

....

.... (2 marks)

Now that you have completed this unit, how confident do you feel?

① Interpreting scatter graphs and correlation

② Using the line of best fit

⑨ Sequences

This unit will help you work with arithmetic sequences.

A01 Fluency check

① Write the next two terms in each sequence.

  a 3, 7, 11, 15, ..., ...   **b** 3.5, 5, 6.5, 8, ..., ...

  c 14, 9, 4, −1, ..., ...   **d** 2, 1.3, 0.6, −0.1, ..., ...

② Evaluate each expression when $n = 5$.

  a $3n =$ ....   **b** $n + 9 =$ ....   **c** $n - 8 =$ ....   **d** $4n + 3 =$ ....

③ **Number sense**

Work out

  a $3 \times 50 + 7 =$ ....   **b** $30 - 4 \times 7 =$ ....   **c** $-5 \times 20 + 35 =$ ....

Key points

Arithmetic sequences are sequences in which the terms increase or decrease by the same number. This number is called the common difference.

The nth term or general term of a sequence is the rule for working out the term at position n.

These **skills boosts** will help you to find the nth term of sequences, use the nth term to generate sequences and decide if a number is in a sequence.

❶ **One-step arithmetic sequences**   ❷ **Two-step arithmetic sequences**

You might have already done some work on sequences. Before starting the first skills boost, rate your confidence using each type of sequence.

①

Work out the nth term of the sequence 2, 4, 6, 8, 10, …

....

....

②

Work out the nth term of the sequence 2, 7, 12, 17, 22, …

....

....

How confident are you?

1 One-step arithmetic sequences

To find the nth term of a sequence, start by finding the common difference between the terms.

Guided practice

Work out the nth term of the sequence 3, 6, 9, 12, 15, …

Work out the common difference.

3   6   9   12   15

+3   +3   …….   …….

n: 1   2   3   …….   …….

Multiply the common difference by n.

$3n$:   3   6   …….   …….   …….

Find the difference between each term and $3n$.

term: 3   6   …….   …….   …….

term − $3n$:   0   0   …….   …….   …….

nth term = …………

If 'term − $3n$' is zero, then the nth term is the common difference multiplied by n.

1. Work out the nth term of each sequence.

   Hint The common difference can be negative.

   a 5, 10, 15, 20, 25, …   …………………………………

   b 9, 18, 27, 36, 45, …   …………………………………

   c 12, 24, 36, 48, 60, …   …………………………………

   d −4, −8, −12, −16, −20, …   …………………………………

   e −7, −14, −21, −28, −35, …   …………………………………

   f $\frac{1}{2}$, 1, $1\frac{1}{2}$, 2, $2\frac{1}{2}$, …   …………………………………

2. Work out the nth term of each sequence.

   a 3, 4, 5, 6, 7, …   …………………………………

   b 8, 9, 10, 11, 12, …   …………………………………

   c $2\frac{1}{2}$, $3\frac{1}{2}$, $4\frac{1}{2}$, $5\frac{1}{2}$, $6\frac{1}{2}$, …   …………………………………

   d 6, 5, 4, 3, 2, …   …………………………………

   e −1, −2, −3, −4, −5, …   …………………………………

   f −4, −3, −2, −1, 0, …   …………………………………

③ Write the first five terms of the sequence with nth term

a $-n$ ....

Hint Substitute $n = 1$ to 5 into the nth term.

b $n + 3$ ....

c $n - 5$ ....

d $4n$ ....

e $-2n$ ....

f $16 - n$ ....

④ Use each nth term to work out the 20th term.

a $2n$ ....

b $n + 5$ ....

c $-n$ ....

d $3 - n$ ....

⑤ Use each nth term to work out the 50th term.

a $3n$ ....

b $n + 6$ ....

c $n - 24$ ....

d $35 - n$ ....

Exam-style question

⑥ Here is a sequence of patterns made with counters.

Pattern number 1   Pattern number 2   Pattern number 3

a Find an expression, in terms of n, for the number of counters in pattern number n. .... **(2 marks)**

b Sian has 75 counters.
She wants to use them to make a pattern in the sequence.
What is the highest number pattern she can make using as many of her counters as possible? .... **(2 marks)**

Reflect How could you check the nth term is correct?

2 Two-step arithmetic sequences

Guided practice

Work out the *n*th term of the sequence 5, 9, 13, 17, 21, ...

Work out the common difference.

5　9　13　17　21

+4　+4　.......　.......

n:　　　1　2　3　......　.......

Multiply the common difference by *n*.

4*n*:　　　4　8　12　......　.......

Find the difference between each term and 4*n*.

term:　　　5　9　......　......　......

term − 4*n*:　1　1　......　......　......

'term − 4*n*' is a constant.
Add this number to the common difference multiplied by *n*.

*n*th term = ...*n* + ...

① Work out the *n*th term of each sequence.

a 2, 6, 10, 14, 18, ...　　　.................................

b 7, 12, 17, 22, 27, ...　　.................................

c 5, 14, 23, 32, 41, ...　　.................................

d 7, 17, 27, 37, 47, ...　　.................................

e 1, 7, 13, 19, 25, ...　　.................................

f −3, 4, 11, 18, 25, ...　　.................................

② Work out the *n*th term of each sequence.

a −1, 1, 3, 5, 7, ...　　　.................................

b −1, −3, −5, −7, −9, ...　.................................

c 1, −2, −5, −8, −11, ...　.................................

d −4, −1, 2, 5, 8, ...　　.................................

e −5, −7, −9, −11, −13, ...　.................................

f −6, −11, −16, −21, −26, ...　.................................

Hint The common difference can be negative.

3 Work out the nth term of each sequence.

    a $3\frac{1}{2}$, 4, $4\frac{1}{2}$, 5, $5\frac{1}{2}$, ...

    b $-4\frac{1}{2}$, -4, $-3\frac{1}{2}$, -3, $-2\frac{1}{2}$, ...

    c $5\frac{1}{2}$, 5, $4\frac{1}{2}$, 4, $3\frac{1}{2}$, ...

    d $\frac{1}{2}$, 0, $-\frac{1}{2}$, -1, $-1\frac{1}{2}$, ...

4 Write the first five terms of the sequence with nth term

    a $2n + 3$ ....

    b $4n - 1$ ....

    c $15 - 2n$ ....

    d $4 - 3n$ ....

5 Use each nth term to work out the 80th term.

    a $2n + 7$ ....

    b $5n - 13$ ....

    c $50 - 3n$ ....

    d $25 - \frac{1}{2}n$ ....

6 **a** Work out the nth term for the sequence 1, 4, 7, 10, 13, ...

    b Is 103 in the sequence? Explain your answer.   ....

        Hint Put 103 equal to the nth term and solve the equation. Is the solution a whole number?

7 **a** Work out the nth term for the sequence 6, 10, 14, 18, ...

    b Is 156 in the sequence? Explain your answer.   ....

Exam-style question

8 The first four terms of a number sequence are 5, 9, 13 and 17.

    a Find the 6th term in this sequence.   .... **(1 mark)**

    b The number 102 is not a term in this sequence. Explain why.

    .... **(2 marks)**

    c Write an expression, in terms of n, for the nth term of this sequence.

    .... **(2 marks)**

Reflect   How could you check whether a given number is in a sequence?

Practise the methods

Answer this question to check where to start.

Check up

Match each sequence with the correct nth term.

A ○ 1, 4, 7, 10, 13, …

B ○ 5, 7, 9, 11, 13, …

C ○ −1, −4, −7, −10, −13, …

D ○ −1, 1, 3, 5, 7, …

E ○ 5, 8, 11, 14, 17, …

F ○ 1, −1, −3, −5, −7, …

1 $2n + 3$ **2** $3n + 2$ **3** $2n - 3$ **4** $3n - 2$ **5** $3 - 2n$ **6** $2 - 3n$

If you matched 1 to B, 2 to E, 3 to D, 4 to A, 5 to F and 6 to C, go to Q2.

If you matched them differently, go to Q1 for more practice.

(1) Work out the nth term of each sequence.

a 7, 14, 21, 28, 35, … ....

b 5, 10, 15, 20, 25, … ....

c 2, 3, 4, 5, 6, … ....

d 6, 5, 4, 3, 2, … ....

e −3, −6, −9, −12, −15, … ....

f −1, 0, 1, 2, 3, … ....

(2) **a** Here are the first five terms of an arithmetic sequence: 6   11   16   21   26

Write, in terms of n, an expression for the nth term of the sequence. ....

b Here are the first five terms of another arithmetic sequence: 9   5   1   −3   −7

Find an expression, in terms of n, for the nth term of this sequence. ....

(3) Here are the first five terms of an arithmetic sequence: 1   8   15   22   29

a Write the next term of this sequence. ....

b Write an expression for the nth term of this sequence. ....

c Work out the 50th term of this sequence. ....

d The number 472 is *not* a term of this sequence. Explain why.

....

Exam-style question

(4) Here are some patterns made from sticks.

Pattern number 1   Pattern number 2   Pattern number 3

a Work out the number of sticks in pattern number n, in terms of n. .... (1 mark)

b Toby says that he will need 70 sticks for pattern number 20.
Is Toby correct? You must give a reason for your answer.

.... (1 mark)

Problem-solve!

1 Here are the first three terms of an arithmetic sequence: 38 32 26
Find the first two terms in the sequence that are less than zero. ....

2 Here are the first four terms of an arithmetic sequence: 2 5 8 11
  a What is the next term in the sequence?
    Explain how you found your answer. .... **(2 marks)**
  b Work out the nth term of the sequence. .... **(1 mark)**
  c Alex says the number 34 is in this sequence.
    Is she correct? You must give a reason for your answer.
    .... **(2 marks)**

3 Here is a sequence
of triangle numbers
made with counters.

Pattern number 1   Pattern number 2   Pattern number 3   Pattern number 4

  a Draw pattern numbers 5 and 6.

  b Write the first ten triangle numbers in this sequence. ....

  c Explain the rule for the sequence of triangle numbers.
    ....

  d Why are triangle numbers not an arithmetic sequence?
    ....

4 The nth term of an arithmetic sequence is $3n + 2$.
Work out whether 93 is a term in this arithmetic sequence.
.... **(2 marks)**

5 **a** Write down the 30th odd number. .... **(1 mark)**
  b The sum of two consecutive odd numbers is 48.
    Find the smaller of these two odd numbers. .... **(2 marks)**
  c Here are the first five terms of an arithmetic sequence: 9 13 17 21 25
    Is 45 a term of this sequence? Show how you got to your answer.
    .... **(3 marks)**

Now that you have completed this unit, how confident do you feel?

1 One-step arithmetic sequences

2 Two-step arithmetic sequences

⑩ Percentages

This unit will help you work out percentages in different situations.

A01 Fluency check

① Work out

a 10% of £140 .... **b** 35% of 260 cm .... **c** 42% of 150 g ....

② Increase £240 by 45%. ....

③ Decrease £186 by 20%. ....

④ **Number sense**

Work out

a 0.2 × 70 .... **b** 0.25 × 220 .... **c** 1.1 × 80 ....

Key points

The original amount is always 100%.

An increase means the new amount will be more; a decrease means the new amount will be less.

These **skills boosts** will help you work out reverse percentages, compound percentages and percentage changes in real-life situations.

① Reverse percentages ② Compound percentages ③ Percentage profit and loss

You might have already done some work on percentages. Before starting the first skills boost, rate your confidence using each method.

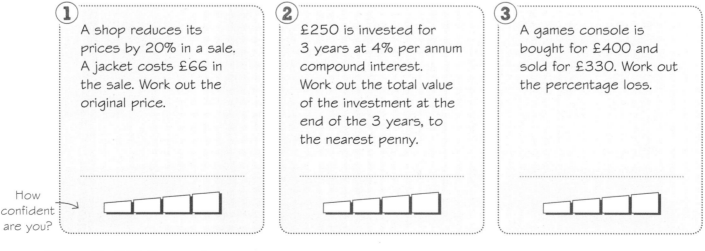

① A shop reduces its prices by 20% in a sale. A jacket costs £66 in the sale. Work out the original price.

② £250 is invested for 3 years at 4% per annum compound interest. Work out the total value of the investment at the end of the 3 years, to the nearest penny.

③ A games console is bought for £400 and sold for £330. Work out the percentage loss.

How confident are you?

1 Reverse percentages

The original amount is always 100%.
If an amount has been increased, the original amount was less than the final amount.
If an amount has been decreased, the original amount was more than the final amount.

Guided practice

Worked exam question

In a sale offering a 20% discount, a pair of trainers costs £68.
What was the original price of the trainers?

Original price of trainers = 100%
Discounted price of trainers = 100% − 20% = 80%
= £68

Work out 1% of the original price.
£68 = 80%   so 1% = £68 ÷ 80 = £...........

Now use 1% to work out the original price.
Original price = 100% = 100 × £......... = £.........

The bar model shows that to work out 1%, you divide £68 by 80.

£68 = 80%   20%

100%

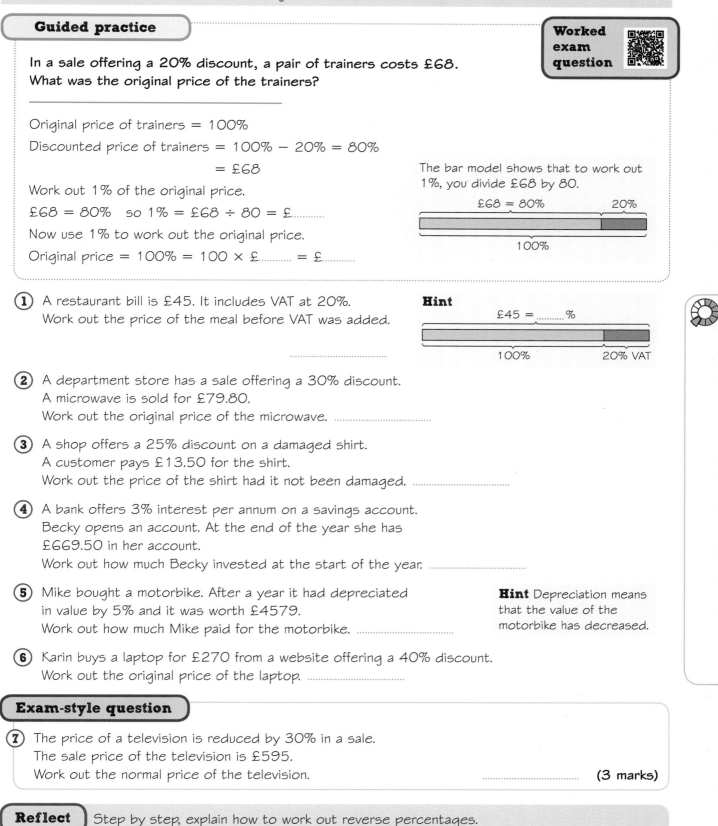

① A restaurant bill is £45. It includes VAT at 20%.
Work out the price of the meal before VAT was added.

....

Hint

£45 = ... %

100%   20% VAT

② A department store has a sale offering a 30% discount.
A microwave is sold for £79.80.
Work out the original price of the microwave. ....

③ A shop offers a 25% discount on a damaged shirt.
A customer pays £13.50 for the shirt.
Work out the price of the shirt had it not been damaged. ....

④ A bank offers 3% interest per annum on a savings account.
Becky opens an account. At the end of the year she has
£669.50 in her account.
Work out how much Becky invested at the start of the year. ....

⑤ Mike bought a motorbike. After a year it had depreciated
in value by 5% and it was worth £4579.
Work out how much Mike paid for the motorbike. ....

Hint Depreciation means
that the value of the
motorbike has decreased.

⑥ Karin buys a laptop for £270 from a website offering a 40% discount.
Work out the original price of the laptop. ....

Exam-style question

⑦ The price of a television is reduced by 30% in a sale.
The sale price of the television is £595.
Work out the normal price of the television. .... **(3 marks)**

Reflect   Step by step, explain how to work out reverse percentages.

2 Compound percentages

A compound percentage is a cumulative percentage. It is calculated by adding/subtracting the percentage increase/decrease at the end of each time period before calculating the next percentage change.

Guided practice

£1500 is invested in a savings account for 3 years at 2.5% per annum compound interest.
Work out the total amount in the savings account at the end of the third year.

Amount at start of first year is £1500.

At the end of the year it will have increased by 2.5%.

Year	Amount at start of year	Calculation	Amount at end of year
1	£1500	£1500 × 1.025	£
2	£1537.50	£1537.50 × 1.025	
3			

Work out the amount at the end of each year.

Use this figure to work out the amount at the end of the next year.

Worked exam question

To increase an amount by 2.5%, multiply by 1.025

Year 2 starts with the total amount from the end of Year 1.
Do not round until you get to your final answer at the end of Year 3.

The calculations in the table show that you multiply £1500 by the multiplier three times:
£1500 × 1.025 × 1.025 × 1.025.
Alternatively, raise the multiplier 1.025 to the power of the number of years:
£1500 × 1.025^3.

① £2400 is invested for 4 years at 3% per annum compound interest.

    a Complete the table to work out the total value of the investment after 4 years.

Year	Amount at start of year	Calculation	Amount at end of year
1	£2400		
2			
3			
4			

    b Work out the value of the investment after 4 years using the multiplier raised to the power of the number of years. ....

② Sally is offered a job with a starting salary of £22 000 and a 5% pay rise each year for the first 3 years. Work out Sally's salary after 3 years. ....

③ The average annual population growth rate for the UK is 0.75%. According to the 2011 census, the population of the UK was 63.2 million (to 1 d.p.).
Work out the predicted population of the UK in 2018.
Give your answer to a suitable degree of accuracy. ....

Hint The answer should be given to the same degree of accuracy as the information given in the question.

Exam-style question

④ A haulage company buys a new fleet of lorries. Each lorry costs £88 000 and depreciates in value by 6% per annum. Work out the value of each lorry after 4 years.
Round your answer to the nearest pound. .... (3 marks)

Reflect
Which method did you prefer in Q1? Why did you prefer this method?

3 Percentage profit and loss

To find percentage profit and loss first convert the profit or loss to a fraction of the original amount.

Worked exam question

Guided practice

A car is bought for £8550 and sold 2 years later for £6900.
Work out the percentage loss.
Give your answer to 1 decimal place (d.p.).

Work out the actual loss.

Actual loss = £8550 − £6900 = £ ....

Express this loss as a fraction of the original amount.

Fractional change = $\dfrac{\text{actual change}}{\text{original amount}}$ = $\dfrac{\text{..........}}{8550}$

> To change from a fraction to a percentage, multiply by 100

Convert the fraction to a percentage.

Percentage change = fractional change × 100 = $\dfrac{\text{..........}}{8550}$ × 100 = ... (to 1 d.p.)

① Abi received a pay rise from £10.25 to £10.50 per hour.
Work out her percentage pay increase.
Give your answer to 1 d.p. ....

Hint First work out the actual reduction in price. Then write the fractional change before converting to a percentage.

② A website reduces the price of a game from £43.99 to £38.99.
Work out the percentage reduction in price.
Give your answer to 3 significant figures (s.f.). ....

③ In 2000, the population of the US was 282.9 million. In 2015, it was 321.8 million.
Work out the percentage increase in the population.
Give your answer to 1 d.p. ....

④ When Zayn was 12 years old, he was 1.42 m tall. Now aged 16 he is 1.78 m tall.
Calculate the percentage increase in Zayn's height.
Give your answer to 1 d.p. ....

⑤ Rosie spends £4500 on shares. A year later she sells her shares for £4265.
Work out Rosie's percentage loss on the sale of her shares.
Give your answer to 3 s.f. ....

⑥ Miriam invests £2600. At the end of 12 months her investment is worth £2690.
Calculate the percentage increase on Miriam's investment.
Give your answer to 3 s.f. ....

Exam-style question

⑦ Shaun's weight decreases from 67.2 kg to 60.1 kg.
Calculate the percentage decrease in Shaun's weight.
Give your answer to 3 s.f. .... (3 marks)

Reflect Without looking at this page, write in your own words how to work out percentage change.

Practise the methods

Answer this question to check where to start.

Check up
£2400 is invested for 4 years at 3% per annum compound interest. Tick the multiplier you would use to find the value of the investment at the end of the fourth year.

A ○ 0.3^4 **B** ○ 0.97^4 **C** ○ 1.03^4 **D** ○ 0.03^4 **E** ○ 0.04^3 **F** ○ 1.04^3

If you ticked C, calculate the value of the investment after 4 years. Then go to Q3.

If you ticked A, B, D, E or F, go to Q1 for more practice.

1. Write the multiplier for

   a an increase of 20% .... **b** an increase of 8% .... **c** a decrease of 42% ....

2. Write the multiplier for

   a £500 invested for 3 years at 5% per annum compound interest ....

   b £850 that depreciates at 7% per annum for 5 years ....

   c £100 invested for 5 years at 3.5% per annum. ....

3. £6000 is invested for 4 years at 2% per annum compound interest.

   Calculate how much the investment will be worth after 4 years. ....

4. Grace buys a gift from a website offering a 35% discount.

   She pays £26.99. Work out the original price of the gift. ....

5. The 2011 census results showed that the population of England was 53 million.
   This was an 8% increase on the previous census in 2001.
   Calculate the population in 2001. Give your answer to 3 s.f. ....

6. A shop reduces the price of a pair of jeans from £58 to £45.
   Work out the percentage discount on the jeans. Give your answer to 1 d.p. ....

7. In order to protect tigers in India, there is a National Conservation Tiger Authority.
   It reported counts of

   • 1411 tigers in 2006   • 1706 tigers in 2011   • 2226 tigers in 2014

   a Work out the percentage increase in tigers from 2006 to 2011. ....

   b Work out the percentage increase in tigers from 2011 to 2014. ....

   Give your answers to 3 s.f.

Exam-style question

8. The value of a new car depreciates by 15% each year for the first 2 years.
   At the end of 2015, the cost of the new car was £17995.
   What will the car be worth at the end of 2017?   ....   (3 marks)

Problem-solve!

1 Steve buys and sells cars.
He has to reach a target of at least 40% profit on each car that he sells.
Steve buys a car for £3500 and sells it for £4950.
Show that Steve has reached his target for this car. .... (3 marks)

2 Jordan invests £3000 in a savings account for 2 years.
The account pays compound interest at an annual rate of
- 1.5% for the first year
- 2.5% for the second year.
How much money will Jordan have in his account at the end of
the 2 years? .... (3 marks)

3 Saira buys a house for £170 000.
She sells the house for £175 100.
Work out Saira's percentage profit. .... (3 marks)

4 A tumble dryer costs £260.
During a sale, its price is reduced to £215.
Work out the percentage reduction in the tumble dryer's price.
Give your answer to 3 s.f. .... (3 marks)

5 On Friday a sandwich bar sells 124 sandwiches.
On Saturday it sells 165 sandwiches.
Work out the percentage increase in the number of sandwiches it sells.
Give your answer to 1 d.p. .... (3 marks)

6 The price of a sofa is reduced by 40% during a sale.
The sale price of the sofa is £1320.
Work out the normal price of the sofa. .... (3 marks)

7 In a sale, prices are reduced by 20%.
A kettle has a sale price of £30.40.
By how much money has the normal price of the kettle
been reduced? .... (3 marks)

8 In a sale, prices are reduced by 30%.
The price of a coat is reduced by £22.50.
Work out the normal price of the coat. .... (3 marks)

Now that you have completed this unit, how confident do you feel?

1 Reverse percentages   **2** Compound percentages   **3** Percentage profit and loss

Answers

Unit 1 Circles

A01 Fluency check

(1) 4.64

(2) 5.8

(3) 6490

(4) **a** $\frac{1}{4}$ **b** $\frac{1}{6}$ **c** $\frac{1}{8}$ **d** $\frac{2}{9}$

(5) **Number sense**

   a 9   **b** 49   **c** 25   **d** 144

Confidence questions

(1) 12.6 cm

(2) 153.9 cm²

(3) 21.4 cm

(4) 39.3 cm²

Skills boost 1 Circumference of a circle

Guided practice

Work out the diameter of the circle.

$d = 2 \times 4.5 = 9$

$C = \pi d$

   $= \pi \times 9$

   $= 28.27433...$

   $= 28.27$ cm (to 2 d.p.)

(1) **a** 21.99 cm   **b** 43.98 mm

   c 113.10 mm   **d** 32.67 cm

(2) **a** 12.57 m   **b** 23.88 mm

   c 41.47 m   **d** 138.23 m

(3) 3.2 m

(4) 8.0 cm

(5) 37.7 mm

Skills boost 2 Area of a circle

Guided practice

$A = \pi r^2$

   $= \pi \times 7^2$

   $= 153.93804$

   $= 153.9$ cm² (to 1 d.p.)

(1) **a** 28.3 cm²   **b** 176.7 mm²

   c 78.5 m²   **d** 18.1 m²

(2) **a** 530.9 cm²   **b** 132.7 m²

   c 1963.5 mm²

(3) **a** 63.6 cm²   **b** 254.5 mm²

   c 16.6 cm²   **d** 1075.2 mm²

(4) 6.2 mm

(5) 1.08 m

Skills boost 3 Arc length and perimeter of a sector

Guided practice

Arc length $= \frac{60}{360}$ of the circumference.

Arc length $= \frac{60}{360} \times \pi d$

$d = 2 \times 4 = 8$ cm

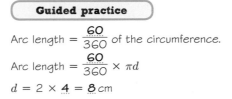

Arc length $= \frac{60}{360} \times \pi \times 8$

       $= 4.18879$

       $= 4.2$ cm (to 1 d.p.)

(1) **a** 9.4 cm   **b** 23.0 mm

(2) **a** 15.42 cm   **b** 7.14 cm

   c 7.71 m   **d** 22.85 cm

(3) **a** 22.7 cm   **b** 8.89 m

   c 30.9 cm   **d** 153 mm

(4) 33.6 cm

Skills boost 4 Area of a sector

Guided practice

$A = \frac{60}{360}$ of the area of the circle.

$A = \frac{60}{360} \times \pi r^2$

   $= \frac{60}{360} \times \pi \times 4^2$

   $= 8.37758$

   $= 8.38$ cm² (to 3 s.f.)

(1) **a** 28.3 cm²   **b** 127 mm²

(2) **a** 25.1 cm²   **b** 19.6 mm²   **c** 0.475 m²   **d** 117 m²

(3) **a** 12.6 cm²   **b** 11.8 m²   **c** 142 cm²   **d** 1760 mm²

(4) 339 cm²

Practise the methods

(1) **a i** 43.98 cm   **ii** 153.94 cm²

   b i 81.68 mm   **ii** 530.93 mm²

   c i 10.05 m   **ii** 8.04 m²

(2) **a** 56.5 cm   **b** 254 cm²

(3) **a** 18.0 cm   **b** 19.2 cm²

(4) **a i** 28.6 cm   **ii** 50.3 cm²

   b i 4.87 m   **ii** 0.873 m²

   c i 24.0 cm   **ii** 34.9 cm²

   d i 42.2 cm   **ii** 86.8 cm²

(5) 23.9 cm

(6) 12.4 cm

Problem-solve!

(1) £53.41   (2) 5.4 cm²

(3) 339 cm²   (4) 15.5 cm²

(5) 15 cm   (6) 3.57 m

Unit 2 Volume and surface area

A01 Fluency check

(1) **a** 32 cm²   **b** 36 cm²   **c** 25 cm²

(2) **a**

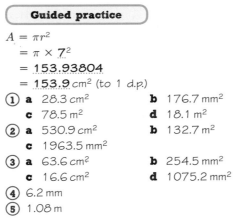

   b

(3) Number sense

   a 42   **b** 216   **c** 480

Confidence questions

(1) Surface area 108 cm², volume 72 cm³

(2) 235 cm³

(3) $h = \dfrac{V}{\pi r^2}$

Skills boost 1 Surface area and volume of a prism

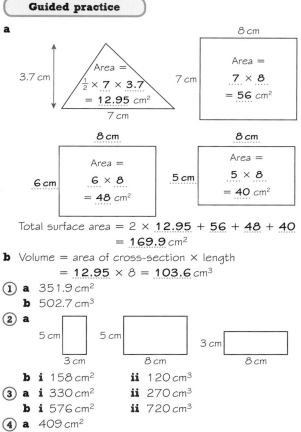

Guided practice

a

Total surface area = 2 × **12.95** + **56** + **48** + **40**
= **169.9** cm²

b Volume = area of cross-section × length
= **12.95** × 8 = **103.6** cm³

(1) **a** 351.9 cm²
   b 502.7 cm³

(2) **a**

   b i 158 cm²   **ii** 120 cm³

(3) **a i** 330 cm²   **ii** 270 cm³
   b i 576 cm²   **ii** 720 cm³

(4) **a** 409 cm²
   b 495 cm³

Skills boost 2 Surface area and volume of more complex solids

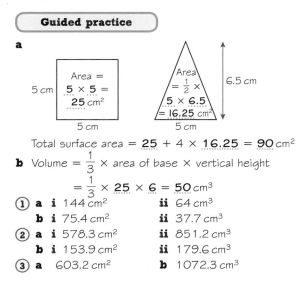

Guided practice

a

Total surface area = **25** + 4 × **16.25** = **90** cm²

b Volume = $\dfrac{1}{3}$ × area of base × vertical height
= $\dfrac{1}{3}$ × **25** × **6** = **50** cm³

(1) **a i** 144 cm²   **ii** 64 cm³
   b i 75.4 cm²   **ii** 37.7 cm³

(2) **a i** 578.3 cm²   **ii** 851.2 cm³
   b i 153.9 cm²   **ii** 179.6 cm³

(3) **a** 603.2 cm²   **b** 1072.3 cm³

Skills boost 3 Length of a prism given its volume

Guided practice

Volume of a cuboid = area of cross-section × **length**
Area of cross-section = 4 × **7** = **28** cm²
Volume = **28** × x
     $x = \dfrac{224}{28} = \textbf{8}$

So the length of the cuboid is **8** cm.

(1) 22.4 cm

(2) 7 cm

(3) **a** 7.5 cm   **b** 12.5 cm   **c** 8 cm

(4) 10.0 cm

Practise the methods

(1) **a**   **b**

   c

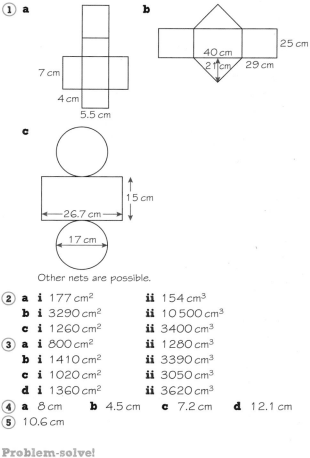

Other nets are possible.

(2) **a i** 177 cm²   **ii** 154 cm³
   b i 3290 cm²   **ii** 10 500 cm³
   c i 1260 cm²   **ii** 3400 cm³

(3) **a i** 800 cm²   **ii** 1280 cm³
   b i 1410 cm²   **ii** 3390 cm³
   c i 1020 cm²   **ii** 3050 cm³
   d i 1360 cm²   **ii** 3620 cm³

(4) **a** 8 cm   **b** 4.5 cm   **c** 7.2 cm   **d** 12.1 cm

(5) 10.6 cm

Problem-solve!

(1) 200 matchboxes

(2) Accept from 104 cm² to 108 cm²

(3) 18.5 cm

(4) 268 cm³

Unit 3 Angles

A01 Fluency check

(1) 180°

(2) **a** 45°   **b** 32°   **c** 76°   **d** 60°

(3) Number sense

   a 83   **b** 156   **c** 108   **d** 1080

Confidence questions

(1) 120°

(2) 360°

(3) 108°

Skills boost 1 Interior and exterior angles

Guided practice

$159 + 83 + p = 180$
$p = 360 - 242 = 118°$

1. **a** $y = 111°$ **b** $y = 87°$ **c** $x = 86°$
   d $x = 131°$ **e** $r = 54°$ **f** $y = 39.5°$
2. **a** 8 sides   **b** $135°$
3. **a** 5 sides   **b** $108°$

Skills boost 2 Angle sum of a polygon

Guided practice

The angle sum of a triangle = $180°$
The pentagon can be split into 3 triangles.
Angle sum of a pentagon = $180 × 3 = 540°$

1. $360°$
2. **a**

Number of sides	Number of triangles	Interior angle sum
3	1	$1 × 180° = 180°$
4	2	$2 × 180° = 360°$
5	3	$3 × 180° = 540°$
6	4	$4 × 180° = 720°$
7	5	$5 × 180° = 900°$
8	6	$6 × 180° = 1080°$

   b Interior angle sum of a polygon
   = (number of sides − 2) × $180°$
3. **a** $2520°$ **b** $3240°$
4. **a** 9 sides   **b** 15 sides

Skills boost 3 Using the angle sum of a polygon

Guided practice

Angle sum = $(n − 2) × 180°$
Angle sum = $4 × 180° = 720°$
Each interior angle = $\frac{720}{6} = 120°$

1. $108°$
2. **a** $135°$ **b** $140°$ **c** $157.5°$
3. **a** $156°$ **b** $126°$ **c** $140°$
4. $135°$

Practise the methods

1. **a**   **b**
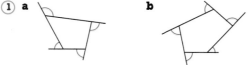

   Accept other correctly drawn exterior angles.
2. $360°$ 3. **a** $41°$ **b** $30°$
4. **a** 9 sides **b** $140°$ 5. $2880°$
6. $144°$ 7. $125°$

Problem-solve!

1. 19 sides   2. $84°$
3. $15°$ 4. $135°$
5. $20°$ 6. $35°$
7. 18 sides

Unit 4 Vectors

A01 Fluency check

1. **a** $8a$ **b** $2b$
   c $3p + 5q$ **d** $-10x + 5y$

2. **Number sense**

a 3   **b** -9 **c** 6   **d** -5
e -15 **f** -72 **g** 60   **h** -24

Confidence questions

1. $\begin{pmatrix} 4 \\ -1 \end{pmatrix}$ 2. $\begin{pmatrix} 4 \\ 6 \end{pmatrix}$ 3. $\begin{pmatrix} 8 \\ 20 \end{pmatrix}$

Skills boost 1 Understand and use vector notation

Guided practice

a $\overrightarrow{AB} = \begin{pmatrix} 3 \\ 2 \end{pmatrix}$ **b** $\overrightarrow{CD} = \begin{pmatrix} -3 \\ -2 \end{pmatrix}$ **c** $\overrightarrow{EF} = \begin{pmatrix} -4 \\ 0 \end{pmatrix}$

1. **a** $\begin{pmatrix} 5 \\ -2 \end{pmatrix}$ **b** $\begin{pmatrix} -5 \\ 2 \end{pmatrix}$
2. **a** $\begin{pmatrix} 6 \\ -4 \end{pmatrix}$ **b** $\begin{pmatrix} 0 \\ -3 \end{pmatrix}$ **c** $\begin{pmatrix} -4 \\ 3 \end{pmatrix}$

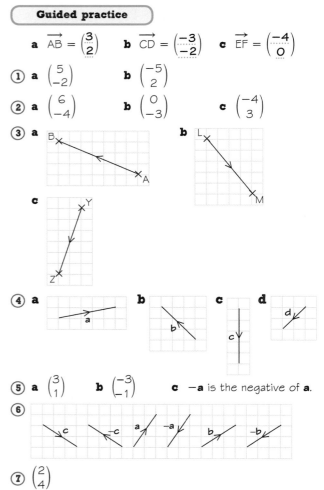

3. **a**   **b**
   c

4. **a**   **b**   **c**   **d**

5. **a** $\begin{pmatrix} 3 \\ 1 \end{pmatrix}$ **b** $\begin{pmatrix} -3 \\ -1 \end{pmatrix}$ **c** $-a$ is the negative of **a**.

6.

7. $\begin{pmatrix} 2 \\ 4 \end{pmatrix}$

Skills boost 2 Add and subtract vectors

Guided practice

$\begin{pmatrix} 4 \\ 1 \end{pmatrix} + \begin{pmatrix} -1 \\ 3 \end{pmatrix} = \begin{pmatrix} 4 + -1 \\ 1 + 3 \end{pmatrix} = \begin{pmatrix} 3 \\ 4 \end{pmatrix}$

1. **a** $\begin{pmatrix} -5 \\ -7 \end{pmatrix}$ **b** $\begin{pmatrix} -5 \\ 4 \end{pmatrix}$ **c** $\begin{pmatrix} -9 \\ 7 \end{pmatrix}$
2. **a** $\begin{pmatrix} 2 \\ 2 \end{pmatrix}$ **b** $\begin{pmatrix} 2 \\ -3 \end{pmatrix}$ **c** $\begin{pmatrix} 4 \\ -1 \end{pmatrix}$

   d **i** $\begin{pmatrix} 4 \\ -1 \end{pmatrix}$ **ii** $x + y = \begin{pmatrix} 2 \\ 2 \end{pmatrix} + \begin{pmatrix} 2 \\ -3 \end{pmatrix} = \begin{pmatrix} 4 \\ -1 \end{pmatrix} = z$

3. **a** $\begin{pmatrix} 1 \\ -6 \end{pmatrix}$ **b**

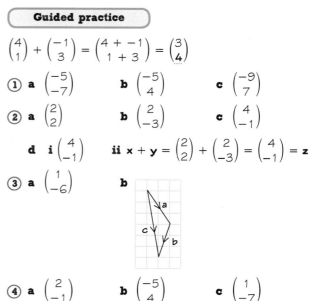

4. **a** $\begin{pmatrix} 2 \\ -1 \end{pmatrix}$ **b** $\begin{pmatrix} -5 \\ 4 \end{pmatrix}$ **c** $\begin{pmatrix} 1 \\ -7 \end{pmatrix}$

(5) **a** $\begin{pmatrix}-2\\-3\end{pmatrix}$ **b**

Skills boost 3 Find multiples of vectors

Guided practice

a $a = \begin{pmatrix}3\\-4\end{pmatrix}$

b $2a = 2 \times \begin{pmatrix}3\\-4\end{pmatrix}$
$= \begin{pmatrix}2 \times 3\\2 \times -4\end{pmatrix}$
$= \begin{pmatrix}6\\-8\end{pmatrix}$

(1) **a** $\begin{pmatrix}0\\8\end{pmatrix}$ **b** $\begin{pmatrix}-6\\15\end{pmatrix}$ **c** $\begin{pmatrix}0\\-12\end{pmatrix}$ **d** $\begin{pmatrix}2\\-5\end{pmatrix}$

(2) **a** $\begin{pmatrix}-2\\4\end{pmatrix}$ **b** $\begin{pmatrix}-6\\9\end{pmatrix}$ **c** $\begin{pmatrix}0\\1\end{pmatrix}$ **d** $\begin{pmatrix}7\\-11\end{pmatrix}$

(3) **a** $\begin{pmatrix}-3\\1\end{pmatrix}$ **b** $\begin{pmatrix}6\\-2\end{pmatrix}$ **c** $\begin{pmatrix}3\\-1\end{pmatrix}$

Practise the methods

(1) **a**   **b**   **c**

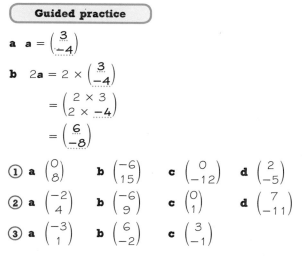

(2)

(3) **a** $\begin{pmatrix}2\\-4\end{pmatrix}$ **b**

(4) **a** $\begin{pmatrix}8\\-3\end{pmatrix}$ **b**

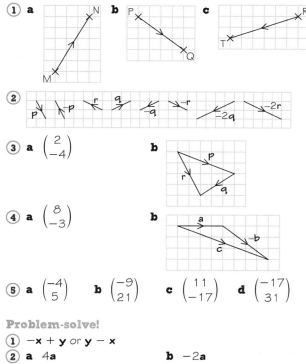

(5) **a** $\begin{pmatrix}-4\\5\end{pmatrix}$ **b** $\begin{pmatrix}-9\\21\end{pmatrix}$ **c** $\begin{pmatrix}11\\-17\end{pmatrix}$ **d** $\begin{pmatrix}-17\\31\end{pmatrix}$

Problem-solve!

(1) $-x + y$ or $y - x$
(2) **a** $4a$ **b** $-2a$
(3) **a** $-a + b$ or $b - a$ **b** $b + a$
(4) **a** $-4b$ **b** $-4b + a$ or $a - 4b$

Unit 5 Transformations
Getting started

AO1 Fluency check

(1) part c only
(2) **a** translation   **b** rotation
 c enlargement

(3) Number sense
 a 4   **b** 18   **c** 15   **d** 6

Confidence questions

(1) 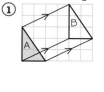   (2)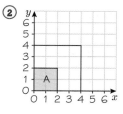

(3) 10 cm
(4) Centre of enlargement (0, 0), scale factor = $\frac{1}{2}$

Skills boost 1 Translations with vectors

Guided practice

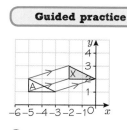

(1)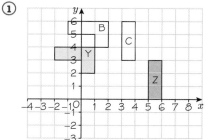

(2) **a–d**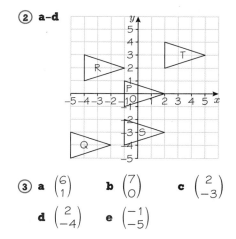

(3) **a** $\begin{pmatrix}6\\1\end{pmatrix}$ **b** $\begin{pmatrix}7\\0\end{pmatrix}$ **c** $\begin{pmatrix}2\\-3\end{pmatrix}$
 d $\begin{pmatrix}2\\-4\end{pmatrix}$ **e** $\begin{pmatrix}-1\\-5\end{pmatrix}$

Skills boost 2 Enlargements with fractional scale factors

Guided practice

The column vector from the centre of enlargement to (6, 2) is $\begin{pmatrix}6\\2\end{pmatrix}$

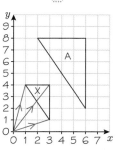

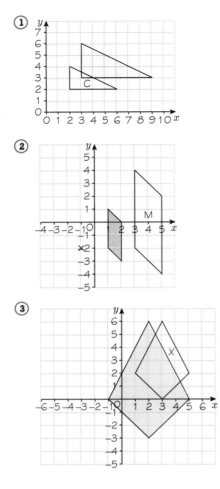

①

②

③

Skills boost 3 Similarity and scale factors

> **Guided practice**

Scale factor $= \frac{3}{2} = 1\frac{1}{2}$

Multiply length AB by the scale factor.

$PQ = 7 \times 1\frac{1}{2} = 10\frac{1}{2}$ cm

The length of rectangle PQRS is $10\frac{1}{2}$ cm.

① **a** WX   **b** WZY   **c** XY

② **a** 2   **b** $\frac{1}{2}$ **c** 2 cm

③ $7\frac{1}{2}$ cm

④ XZ = 9 cm, YZ = 6 cm

Skills boost 4 Describing enlargements with fractional scale factors

> **Guided practice**

Scale factor $= \frac{\text{side of shape B}}{\text{side of shape A}} = \frac{2}{4} = \frac{1}{2}$

The transformation is an enlargement with scale factor $\frac{1}{2}$ with centre of enlargement (8, 9).

① **a** $\frac{1}{2}$ **b** $\frac{1}{3}$ **c** $1\frac{1}{2}$

② **a** enlargement with scale factor $\frac{1}{2}$ from centre of enlargement (0, 1)

    b enlargement with scale factor $\frac{1}{4}$ from centre of enlargement (6, 1)

① **a**   **b**   **c**   **d**

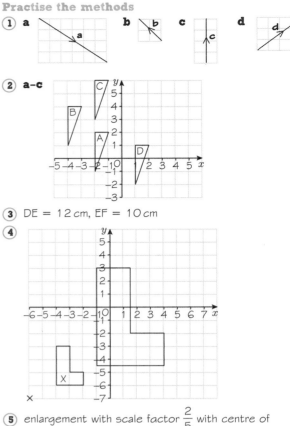

② **a-c**

③ DE = 12 cm, EF = 10 cm

④

⑤ enlargement with scale factor $\frac{2}{5}$ with centre of enlargement (−6, 5)

Problem-solve!

① 15 cm

② $28\frac{1}{2}$ cm

③ 3 cm²

④ **a, b**

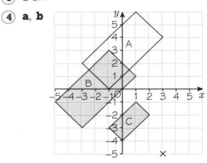

⑤ $\begin{pmatrix} -a \\ -b \end{pmatrix}$

Unit 6 Averages

Get started

> **A01 Fluency check**

① **a** 32.2   **b** 33   **c** 33

② **a** 3.5   **b** 42.73   **c** 14.0

> **③ Number sense**

    a 165   **b** 17.5   **c** 99

Confidence questions

① 38 years (to the nearest whole number)

② Extreme values may make it not representative.

③ 9

Skills boost 1 Estimating the mean from a grouped frequency table

Height, h (cm)	Frequency	Midpoint of class	What the information means	Midpoint × frequency
$0 < h \leqslant 10$	3	5	Assume three plants each measure 5 cm	$3 \times 5 = 15$
$10 < h \leqslant 20$	20	15	Assume 20 plants each measure 15 cm	$20 \times 15 = 300$
$20 < h \leqslant 30$	52	25	Assume 52 plants each measure 25 cm	$52 \times 25 = 1300$
$30 < h \leqslant 40$	77	35	Assume 77 plants each measure 35 cm	$77 \times 35 = 2695$
$40 < h \leqslant 50$	23	45	Assume 23 plants each measure 45 cm	$23 \times 45 = 1035$
Total	175			5345

Estimated total height of all bean plants = 5345
Total number of bean plants = 175
Estimate of mean = 5345 ÷ 175 = 30.5
(1) 0.88 seconds   (2) 1.70 m

Skills boost 2 Which average is best?

Disadvantage of the mean: the mean is affected by the extreme value of £187 000.
Advantage of the median: the median is not affected by the extreme value of £187 000 and it would not change if any salary other than £45 000 changes.
Disadvantage of the mode: there is not a mode as all the salaries are different.

(1)

Average	Advantages	Disadvantages
Mean	Every value makes a difference	Affected by extreme values
Median	Not affected by extreme values, may not change if a value changes	Does not use every value
Mode	Not affected by extreme values, easy to find, can be used with non-numerical data	There may not be a mode

(2) The mean would be better for data set B as data set A has the extreme value of 11.2.
(3) The median works for both data sets but is particularly useful for data set A as it has the extreme value of 11.2.
(4) The mode would be better for data set B as data set A does not have a mode. Also, only the mode can be used for non-numerical data.
(5) a Mean – there are no extreme values, there is no mode, the median would work but doesn't take into account every number.
   b Mode – as the type of pet is not a numerical value.
   c Median – there is no mode and there is an extreme value of 21.8.
   d Any – all averages work well and give similar answers.

(6) a 5.7 minutes
   b The mean is affected by the extreme value of 27 minutes.
   c The median would the most appropriate average as there are two modes and the median is not affected by extreme values.

Skills boost 3 Calculating missing data values

Mean = total ÷ how many numbers
  10 = total ÷ 5
Total = 50
  ? = 50 − 38 = 12
The first number is 12
(1) 90
(2) 22
(3) 147
(4) Students' own answers: check that the five numbers are different and total 55.
(5) Students' own answers: check that two of the numbers are the same and the other four are different. All six numbers must total 90.
(6) 7, 8 or 9
(7) Students' own answers: check the seven numbers have a mode of 12, median of 12 and range of 8.

Practise the methods
(1) a 22.5   b 77.5   c 16.5
(2) 3.32 kg
(3) a Mode – only the mode can be used for non-numerical data.
   b Median – the data set is bi-modal and there is an extreme value of 125.
   c Mean – there are no extreme values, there is no mode, the median would work but doesn't take into account every number.
(4) Students' own answers: check that the six numbers are different and total 84.
(5) a 22   b 13

Problem-solve!
(1) 8   (2) 7, 7, 10   (3) 84.4
(4) 68   (5) 12.9 minutes

Unit 7 Probability

① **a** 1, 4, 9, 16, 25, 36, 49, 64
b 2, 3, 5, 7, 11, 13, 17, 19
c 1, 2, 3, 4, 6, 12

② **Number sense**

a $\frac{3}{8}$ **b** $\frac{5}{14}$ **c** $1\frac{1}{5}$ **d** $\frac{8}{9}$

Confidence questions

① 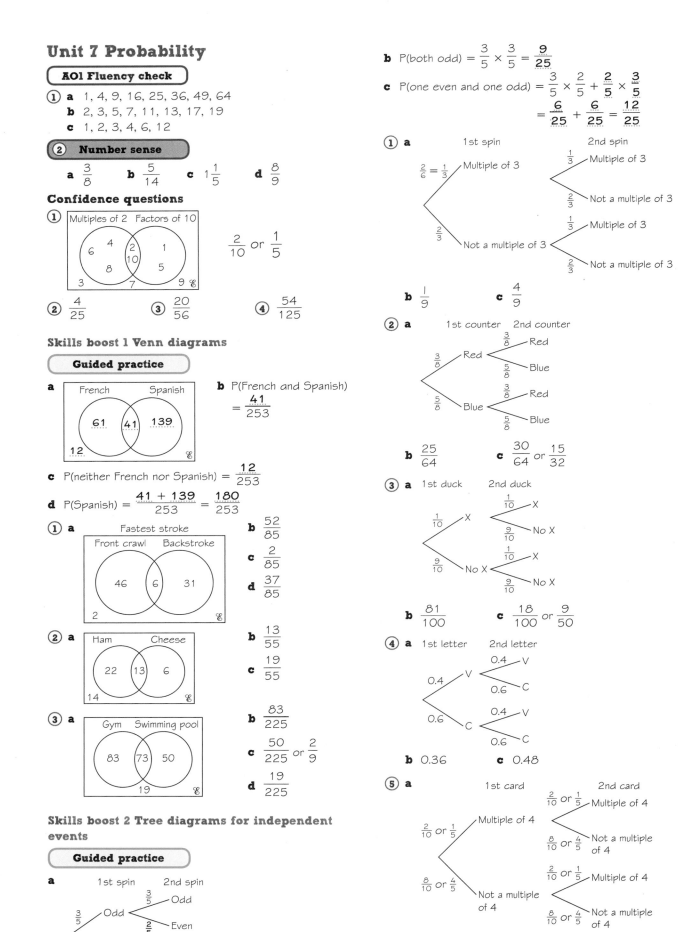 $\frac{2}{10}$ or $\frac{1}{5}$

② $\frac{4}{25}$ ③ $\frac{20}{56}$ ④ $\frac{54}{125}$

Skills boost 1 Venn diagrams

Guided practice

a

b P(French and Spanish) $= \frac{41}{253}$

c P(neither French nor Spanish) $= \frac{12}{253}$

d P(Spanish) $= \frac{41 + 139}{253} = \frac{180}{253}$

① **a** Fastest stroke

b $\frac{52}{85}$

c $\frac{2}{85}$

d $\frac{37}{85}$

② **a**

b $\frac{13}{55}$

c $\frac{19}{55}$

③ **a**

b $\frac{83}{225}$

c $\frac{50}{225}$ or $\frac{2}{9}$

d $\frac{19}{225}$

Skills boost 2 Tree diagrams for independent events

Guided practice

a

b P(both odd) $= \frac{3}{5} \times \frac{3}{5} = \frac{9}{25}$

c P(one even and one odd) $= \frac{3}{5} \times \frac{2}{5} + \frac{2}{5} \times \frac{3}{5}$
$= \frac{6}{25} + \frac{6}{25} = \frac{12}{25}$

① **a**

b $\frac{1}{9}$ **c** $\frac{4}{9}$

② **a**

b $\frac{25}{64}$ **c** $\frac{30}{64}$ or $\frac{15}{32}$

③ **a**

b $\frac{81}{100}$ **c** $\frac{18}{100}$ or $\frac{9}{50}$

④ **a**

b 0.36   **c** 0.48

⑤ **a**

b $\frac{16}{100}$ or $\frac{4}{25}$ **c** $\frac{32}{100}$ or $\frac{8}{25}$

Skills boost 3 Tree diagrams for dependent events

Guided practice

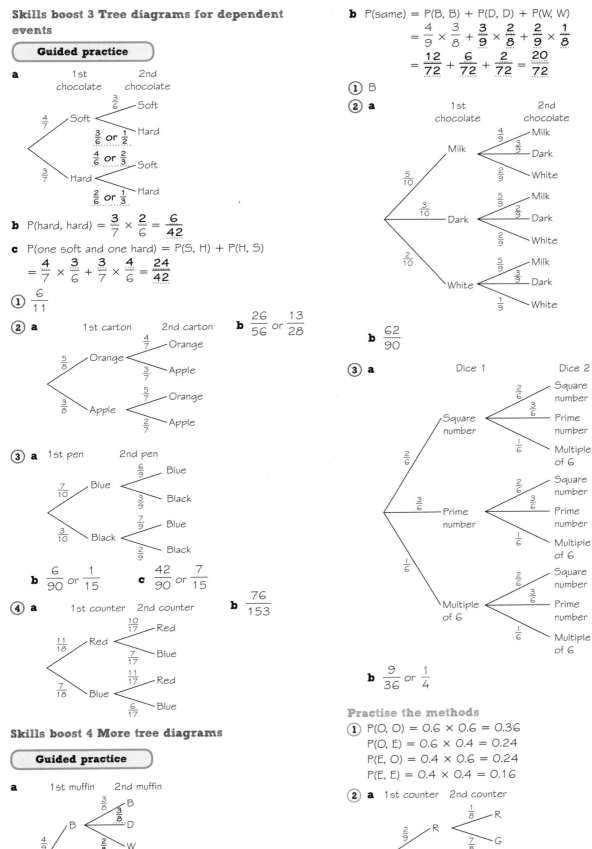

a

1st chocolate — 2nd chocolate

$\frac{4}{7}$ Soft

$\frac{3}{6}$ Soft

$\frac{3}{6}$ or $\frac{1}{2}$ Hard

$\frac{4}{6}$ or $\frac{2}{3}$ Soft

$\frac{3}{7}$ Hard

$\frac{2}{6}$ or $\frac{1}{3}$ Hard

b $P(\text{hard, hard}) = \frac{3}{7} \times \frac{2}{6} = \frac{6}{42}$

c $P(\text{one soft and one hard}) = P(S, H) + P(H, S)$

$= \frac{4}{7} \times \frac{3}{6} + \frac{3}{7} \times \frac{4}{6} = \frac{24}{42}$

① $\frac{6}{11}$

② **a**

1st carton — 2nd carton   **b** $\frac{26}{56}$ or $\frac{13}{28}$

$\frac{5}{8}$ Orange

$\frac{4}{7}$ Orange

$\frac{3}{7}$ Apple

$\frac{3}{8}$ Apple

$\frac{5}{7}$ Orange

$\frac{2}{7}$ Apple

③ **a** 1st pen — 2nd pen

$\frac{7}{10}$ Blue

$\frac{6}{9}$ Blue

$\frac{3}{9}$ Black

$\frac{3}{10}$ Black

$\frac{7}{9}$ Blue

$\frac{2}{9}$ Black

b $\frac{6}{90}$ or $\frac{1}{15}$ **c** $\frac{42}{90}$ or $\frac{7}{15}$

④ **a** 1st counter   2nd counter   **b** $\frac{76}{153}$

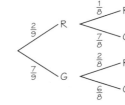

$\frac{11}{18}$ Red

$\frac{10}{17}$ Red

$\frac{7}{17}$ Blue

$\frac{7}{18}$ Blue

$\frac{11}{17}$ Red

$\frac{6}{17}$ Blue

Skills boost 4 More tree diagrams

Guided practice

a   1st muffin   2nd muffin

$\frac{4}{9}$ B

$\frac{3}{8}$ B

$\frac{3}{8}$ D

$\frac{2}{8}$ W

$\frac{3}{9}$ D

$\frac{4}{8}$ B

$\frac{2}{8}$ D

$\frac{2}{8}$ W

$\frac{2}{9}$ W

$\frac{4}{8}$ B

$\frac{3}{8}$ D

$\frac{1}{8}$ W

b $P(\text{same}) = P(B, B) + P(D, D) + P(W, W)$

$= \frac{4}{9} \times \frac{3}{8} + \frac{3}{9} \times \frac{2}{8} + \frac{2}{9} \times \frac{1}{8}$

$= \frac{12}{72} + \frac{6}{72} + \frac{2}{72} = \frac{20}{72}$

① B

② **a**

1st chocolate — 2nd chocolate

$\frac{5}{10}$ Milk

$\frac{4}{9}$ Milk

$\frac{3}{9}$ Dark

$\frac{2}{9}$ White

$\frac{3}{10}$ Dark

$\frac{5}{9}$ Milk

$\frac{2}{9}$ Dark

$\frac{2}{9}$ White

$\frac{2}{10}$ White

$\frac{5}{9}$ Milk

$\frac{3}{9}$ Dark

$\frac{1}{9}$ White

b $\frac{62}{90}$

③ **a**   Dice 1   Dice 2

$\frac{2}{6}$ Square number

$\frac{2}{6}$ Square number

$\frac{3}{6}$ Prime number

$\frac{1}{6}$ Multiple of 6

$\frac{3}{6}$ Prime number

$\frac{2}{6}$ Square number

$\frac{3}{6}$ Prime number

$\frac{1}{6}$ Multiple of 6

$\frac{1}{6}$ Multiple of 6

$\frac{2}{6}$ Square number

$\frac{3}{6}$ Prime number

$\frac{1}{6}$ Multiple of 6

b $\frac{9}{36}$ or $\frac{1}{4}$

Practise the methods

① $P(O, O) = 0.6 \times 0.6 = 0.36$

$P(O, E) = 0.6 \times 0.4 = 0.24$

$P(E, O) = 0.4 \times 0.6 = 0.24$

$P(E, E) = 0.4 \times 0.4 = 0.16$

② **a** 1st counter   2nd counter

$\frac{2}{9}$ R

$\frac{1}{8}$ R

$\frac{7}{8}$ G

$\frac{7}{9}$ G

$\frac{2}{8}$ R

$\frac{6}{8}$ G

b $\frac{42}{72}$

c $\frac{28}{72}$

3 a

 1st spin   2nd spin   **b** 0.38   **c** 0.62

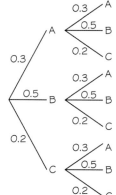

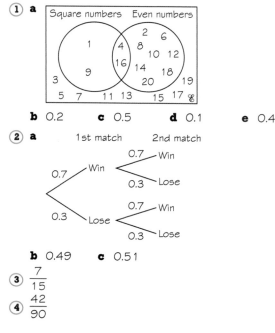

Problem-solve!

1 a

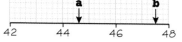

Square numbers   Even numbers

1

4   8   6
16 10 12
9   14 18
        20
3
5 7   11 13   15 17 19

b 0.2   **c** 0.5   **d** 0.1   **e** 0.4

2 a   1st match   2nd match

Win
0.7   Win
        0.7   Win
        0.3   Lose

0.3   Lose
        0.7   Win
        0.3   Lose

b 0.49   **c** 0.51

3 $\frac{7}{15}$

4 $\frac{42}{90}$

Unit 8 Scatter graphs

AO1 Fluency check

1 a Student's own graph showing positive correlation
b Student's own graph showing negative correlation
c Student's own graph showing no correlation

2 Number sense

42   44   46   48

Confidence questions

1 Positive correlation
2 The tree's height is outside the range of his plotted data.

Skills boost 1 Interpreting scatter graphs and correlation

Guided practice

a As the length of the birds **increases**, the wingspan **increases**.
b Positive correlation

1 a As the number of weeks spent in the top 10 of the singles chart increases, the number of weeks spent in the top 10 of the albums chart increases.
b Positive correlation

2 a As the engine size increases, the time it takes to go from 0 to 60 mph decreases.
b Negative correlation

3 a

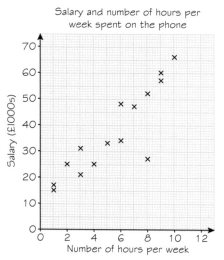

Salary and number of hours per week spent on the phone

b As the salary increases the number of hours spent on the phone increases. The scatter graph shows positive correlation between the two variables.
c The relationship between the two variables is not causal. It is unlikely that a higher salary *causes* the time a person spends on the phone to increase.

Skills boost 2 Using the line of best fit

Guided practice

a

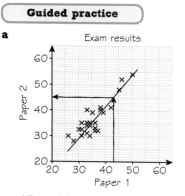

Exam results

45 or 46 marks

b

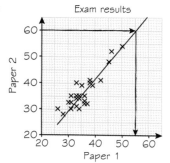

Exam results

approx 55 marks

c The estimate for part **a** is more reliable because **it is within the range of the given data.**

1 a 200 g/km   **b** 129 g/km
c Part b as **it is within the range of the given data.**

Practise the methods

(1) **a** B, negative correlation

    b C, no correlation

    c A, positive correlation

(2) **a** As the cost of the cars increases, the top speed increases.

    b Positive correlation

    c It is not within range of the given data.

(3) **a** 27 or 28 marks

    b 58 marks

    c Seema's mark as it is within the range of the given data.

Problem-solve!

(1) **a** As wealth increases, the life expectancy increases.

    b approx. 77.4 years

    c Age 90 is not a reliable estimate for the life expectancy as it is outside the range of given data.

(2) **a**

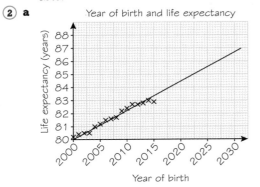

Year of birth and life expectancy

    b You cannot assume a linear relationship. It is outside the range of the given data; extrapolation does not give a reliable estimate.

Unit 9 Sequences

AO1 Fluency check

(1) **a** 19, 23   **b** 9.5, 11

    c $-6, -11$ **d** $-0.8, -1.5$

(2) **a** 15   **b** 14   **c** -3 **d** 23

(3) Number sense

    a 157   **b** 2   **c** -65

Confidence questions

(1) $2n$

(2) $5n - 3$

Skills boost 1 One-step arithmetic sequences

Guided practice

Work out the common difference.

3   6   9   12   15

+3   +3   **+3**   **+3**

n: 1   2   3   **4**   **5**

$3n$:   3   6   **9**   **12**   **15**

term: 3   6   **9**   **12**   **15**

term $- 3n$: 0   0   **0**   **0**   **0**

nth term $= \mathbf{3n}$

(1) **a** $5n$ **b** $9n$ **c** $12n$

    d $-4n$ **e** $-7n$ **f** $\frac{1}{2}n$

---

(2) **a** $n + 2$ **b** $n + 7$ **c** $n + 1\frac{1}{2}$

    d $7 - n$ **e** $-n$ **f** $n - 5$

(3) **a** $-1, -2, -3, -4, -5$

    b 4, 5, 6, 7, 8

    c $-4, -3, -2, -1, 0$

    d 4, 8, 12, 16, 20

    e $-2, -4, -6, -8, -10$

    f 15, 14, 13, 12, 11

(4) **a** 40   **b** 25   **c** -20 **d** -17

(5) **a** 150   **b** 56   **c** 26   **d** -15

(6) **a** $2n$

    b pattern number 37

Skills boost 2 Two-step arithmetic sequences

Guided practice

Work out the common difference.

5   9   13   17   21

+4   +4   **+4**   **+4**

n: 1   2   3   **4**   **5**

$4n$:   4   8   12   **16**   **20**

term: 5   9   **13**   **17**   **21**

term $- 4n$: 1   1   **1**   **1**   **1**

nth term $= \mathbf{4n + 1}$

(1) **a** $4n - 2$ **b** $5n + 2$ **c** $9n - 4$

    d $10n - 3$ **e** $6n - 5$ **f** $7n - 10$

(2) **a** $2n - 3$ **b** $1 - 2n$ **c** $4 - 3n$

    d $3n - 7$ **e** $-2n - 3$ **f** $-5n - 1$

(3) **a** $\frac{1}{2}n + 3$ **b** $\frac{1}{2}n - 5$

    c $6 - \frac{1}{2}n$ **d** $1 - \frac{1}{2}n$

(4) **a** 5, 7, 9, 11, 13   **b** 3, 7, 11, 15, 19

    c 13, 11, 9, 7, 5   **d** $1, -2, -5, -8, -11$

(5) **a** 167   **b** 387   **c** -190 **d** -15

(6) **a** $3n - 2$

    b Yes, when $3n - 2 = 103$, $n = 35$

(7) **a** $4n + 2$

    b No, when $4n + 2 = 156$, n is not a whole number.

(8) **a** 25

    b All the terms in the sequence are odd numbers and 102 is an even number, so 102 cannot be in the sequence.

    c $4n + 1$

Practise the methods

(1) **a** $7n$ **b** $5n$ **c** $n + 1$

    d $7 - n$ **e** $-3n$ **f** $n - 2$

(2) **a** $5n + 1$ **b** $13 - 4n$

(3) **a** 36   **b** $7n - 6$ **c** 344

    d The nth term is $7n - 6$ and when $7n - 6 = 472$, n is not a whole number. Therefore 472 is not a term in the sequence.

(4) **a** $3n + 1$

    b The 20th term is $3 \times 20 + 1 = 61$ so Toby will need 61 sticks, not 70.

Problem-solve!

(1) -4 and -10

(2) **a** 14

    The rule is to add 3 each time, so the next term is $11 + 3 = 14$.

    b $3n - 1$

c No, when $3n - 1 = 34$, n is not a whole number so 34 is not in the sequence

3 a

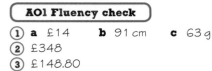

Pattern number 5   Pattern number 6

b 1, 3, 6, 10, 15, 21, 28, 36, 45, 55

c The rule is to add one extra each time, so +1, +2, +3, +4, etc.

d The rule does not add the same amount each time (there is not a common difference).

4 No, when $3n + 2 = 93$, n is not a whole number.

5 a 59

b 23

c nth term is $4n + 5$ and when $4n + 5 = 45$, $n = 10$, so 45 is in the sequence (10th term).

Unit 10 Percentages

AO1 Fluency check

1 a £14   **b** 91 cm   **c** 63 g

2 £348

3 £148.80

4 Number sense

a 14   **b** 55   **c** 88

Confidence check

1 £82.50

2 £281.22

3 17.5%

Skills boost 1 Reverse percentages

Guided practice

£68 = 80% so 1% = £68 ÷ 80 = **£0.85**
Original price = 100% = 100 × £0.85 = **£85**

1 £37.50   **2** £114

3 £18   **4** £650

5 £4820   **6** £450

7 £850

Skills boost 2 Compound percentages

Guided practice

Year	Amount at start of year	Calculation	Amount at end of year
1	£1500	£1500 × 1.025	£1537.50
2	£1537.50	£1537.50 × 1.025	£1575.9375
3	£1575.9375	£1575.9375 × 1.025	£1615.3359

The total amount after the third year is £1615.34

1 a

Year	Amount at start of year	Calculation	Amount at end of year
1	£2400	£2400 × 1.03	£2472
2	£2472	£2472 × 1.03	£2546.16
3	£2546.16	£2546.16 × 1.03	£2622.5448
4	£2622.5448	£2622.5448 × 1.03	£2701.22

b £2701.22

2 £25 467.75

3 66.6 million people

4 £68 706

Skills boost 3 Percentage profit and loss

Guided practice

Actual loss = £8550 − £6900 = **£1650**

$$\text{Fractional change} = \frac{\text{actual change}}{\text{original amount}} = \frac{1650}{8550}$$

Percentage change = fractional change × 100

$$= \frac{1650}{8550} \times 100 = \underline{19.3} \text{ (to 1 d.p.)}$$

1 2.4%   **2** 11.4%

3 13.8%   **4** 25.4%

5 5.22%   **6** 3.46%

7 10.6%

Practise the methods

1 a 1.2   **b** 1.08   **c** 0.58

2 a $1.05^3 = 1.157625$

b $0.93^5 = 0.6957$ (to 4 d.p.)

c $1.035^5 = 1.1877$ (to 4 d.p.)

3 £6494.59

4 £41.52

5 49.1 million people.

6 22.4%

7 a 20.9%

b 30.5%

8 £13 001.39

Problem-solve!

1 Actual profit = £1450

$$\text{Fraction profit} = \frac{1450}{3500}$$

Percentage profit = 41.4%, which is greater than 40%.

Steve reaches his target.

2 £3121.13   **3** 3%

4 17.3%   **5** 33.1%

6 £2200   **7** £7.60

8 £75